The Quartz Mill Operator's Handbook

by P.M. Randall

with an introduction by Kerby Jackson

Introduction

It has been years since P.M. Randall released his important publication "The Quartz Operator's Handbook". First released in 1871, this important volume has now been out of print for years and has been unavailable to the mining community since those days, with the exception of expensive original collector's copies and poorly produced digital editions.

It has often been said that "*gold is where you find it*", but even beginning prospectors understand that their chances for finding something of value in the earth or in the streams of the Golden West are dramatically increased by going back to those places where gold and other minerals were once mined by our forerunners. Despite this, much of the contemporary information on local mining history that is currently available is mostly a result of mere local folklore and persistent rumors of major strikes, the details and facts of which, have long been distorted. Long gone are the old timers and with them, the days of first hand knowledge of the mines of the area and how they operated. Also long gone are most of their notes, their assay reports, their mine maps and personal scrapbooks, along with most of the surveys and reports that were performed for them by private and government geologists. Even published books such as this one are often retired to the local landfill or backyard burn pile by the descendents of those old timers and disappear at an alarming rate. Despite the fact that we live in the so-called "Information Age" where information is supposedly only the push of a button on a keyboard away, true insight into mining properties remains illusive and hard to come by, even to those of us who seek out this sort of information as if our lives depend upon it. Without this type of information readily available to the average independent miner, there is little hope that our metal mining industry will ever recover.

This important volume and others like it, are being presented in their entirety again, in the hope that the average prospector will no longer stumble through the overgrown hills and the tailing strewn creeks without being well informed enough to have a chance to succeed at his ventures.

Kerby Jackson
Josephine County, Oregon
May 2016

CONTENTS.

CONTENTS.

PREFACE.

In the authorship of the first edition of the "Quartz Operator's Hand-Book," published in San Francisco, A. D. 1865, the undersigned, through courtesy, associated with his own, the name of Mr. Zenas Wheeler, his partner at the time in mining engineering, design and structure of mining machinery, and the practical treatment of ores. But in consideration of the decease of Mr. Wheeler, and the extent of the changes made in the revision of the "Hand-Book," the writer has not felt at liberty to further connect with its authorship the name of his lamented friend, of whom, as a just tribute to his memory it must be said, that no one did more toward perfecting the quartz-mining machinery in most general use on the Pacific coast; corroborative of the truth of which, reference need only be had to his inventions of the High Mortar, Gib-Tappet, Self-Regulating Quicksilver Discharge Apparatus, Continuous Working Concentrator and Grinder, and Amalgamator; also, to his associate inventions of the Tractory Grinders and Amalgamators, and Conoidal Separator. As the conclusions of one so eminently qualified as Mr. Wheeler to judge of the relative merits of quartz-mining machinery are of the highest practical importance, it seems due to all concerned, to state that the Straight Battery, Excelsior Grinder and Amalgamator, and Conoidal Separator, described in this work, are the only machines which, of their respective kinds, had his entire approval.

The principal changes referred to above consist in striking out of the "Hand-Book" the more abstruse matter, and in adding that which is regarded at this time more directly applicable to the wants of the mining public.

The object of the author, in the preparation of the work in hand, has been to present a clear and comprehensive exposition of mineral veins, and the means and modes chiefly employed for the mining and working of their ores—more especially those containing gold and silver.

To secure this end, the standard authors, the most skillful miners, mechanics, millmen and mill-wrights have been freely and closely consulted.

The consequent deductions have been carefully compared with the writer's own experience for many years in working mineral veins and in treating their ores and metals. The best authenticated results therefrom have been adopted and set down with a view to systematic arrangement, conciseness and perspicuity. For the articles on the "Examination of Minerals," and the "Behavior of Solutions of Metallic Oxides with Re-Agents," due acknowledgments of indebtedness are here made respectively to Dr. Lamborne's "Metallurgy of Silver and Lead," and to "Noad's Chemical Analysis."

To what extent the author has succeeded in his object, is submitted with no little diffidence to the decision of the public.

P. M. RANDALL.

NEW YORK, January, 1871.

QUARTZ OPERATOR'S HAND-BOOK.

MINERAL VEINS.

MINERAL VEINS are masses of rock, usually of the tabular form intersecting other rocks.

One portion of a mineral vein is called "gangue matrix" or vein-stone, consisting of quartz, calcite, barytes, fluor, etc.; the other portion—ore, consisting of a metal compounded with some other substance as oxygen, sulphur or carbon.

The term "ore" is also employed, sometimes to signify a *native* metal, and sometimes the entire material of a vein.

Mineral Veins occur both in stratified and unstratified rocks, also between the two. Those in the stratified generally cross the strata, but sometimes run parallel with the layers. When a vein is not vertical, the adjacent rocks on the upper side are termed its "hanging wall," and those on the lower its "foot-wall." Usually, however, between a vein and one of its walls proper is a clayey seam called "selvage," or, more familiarly, "gouge."

A layer of half decomposed rock adjoining a vein is

called "flucan." The "flucan" generally accompanies cross courses and slides.

Veins differ greatly as to thickness—some averaging a few inches, and others several feet. They seldom have for any considerable distance their opposite walls quite parallel, but alternately swell and contract, and sometimes *pinch out* entirely.

The spaces thus wanting in ore are termed "faults." When these occur the walls sometimes come nearly or quite together, and sometimes are kept apart by the " gouge."

The gouge is greatly relied on by the miner to lead him to the vein when it has been displaced. The term " fault," however, is used in a wider sense, to signify any displacement or interruption in a vein's continuity.

When veins are very thick they usually contain much earthy matter and fragments of rock mingled with the ore. In veins of this description, also in the more swollen portions of irregular veins, are occasionally found imbedded large blocks of detached wall-rock and bowlder, like masses of barren quartz rock, which respectively are known to miners by the names of " horses," " riders " and " white horses."

The principal mineral veins extend downward to unknown depths, and horizontally, in many instances a distance of many miles.

As a general rule, a proper vein increases in thickness with its descent into the earth's crust, thus presenting the form of a wedge having its edge upward.

Those portions of a vein which rise above the surface of the earth are called "croppings," "outcrops," and "blossoms."

The inclination with which a vein penetrates the earth is termed its "dip," and is measured by the angle which it makes with the horizon. The "dip" varies in different veins from an angle of a few degrees to that of a right angle. The direction in which a vein runs horizontally is called its "bearing" or "strike," and for the most part is parallel with the mountain range in which it is found.

No general law, however, seems to obtain as to the direction of veins in different mining districts. Thus, when carrying the same metal in kind, they run in some districts north and south, and in others east and west. In the same district different veins containing the same metal in kind usually run parallel, but when containing each a different metal, they frequently run in divers directions.

In Cornwall, England, for example, the tin and copper veins run east and west, while the lead veins run north and south.

SPECIES OF MINERAL VEINS.

Mineral veins present a variety of forms, and hence in some localities are distinguished into species, of which the principal are the *Rake, Pipe, Flat or Dilated, Interlaced and Accumulated.*

1st. **The Rake Vein** is of unknown depths, and often many miles in length. Commencing at the surface of the earth it cuts the strata downward, nearly perpendicular to their plane of stratification. For the most part it evidently also intersects the unstratified rocks, and probably passes entirely through the earth's crust to the central mass. The "dip" in different rake veins varies from an inclination of a few degrees to a perpendicular. This class embraces the large proportion of mineral veins proper and is the one most valued by operators in mines.

2d. **The Pipe Vein** runs forward end-wise in a hole in the earth's crust. It resembles an irregular shaft filled with mineral substance, rather than a tabular mass of ore. It generally does not traverse the strata, but lies between the layers. Its "dip" is usually less than that of the rake vein. Veins of this class vary greatly in size, some being only a few feet in diameter and others several yards.

3d. **The Flat or Dilated Vein** is a flattened mass of ore between the strata or beds of other rocks. Its relations to the adjacent formations are much the same as those of a stratum of coal to its inclosures. Its position, also like that of a stratum of coal, is usually horizontal or not much inclined. But when breaks or other interruptions occur in a dilated vein, it differs from a coal seam by greatly varying in thickness within the limits of a few feet. At these breaks or slips often occur the richer deposits of metallic substances. Similar to the rake vein, the dilated in its course is subject to contractions and expansions.

4th. **The Interlaced Vein** is composed of numerous small veins running together in the form of irregular net-work. Different members of the same vein, or system of veins, though running in the same general direction, often vary considerably from each other in " dip."

Interlaced veins are chiefly enclosed by rocks of primitive formation, and more frequently bear the ores of tin than those of any other metal.

5th. **The Accumulated Vein** is a large, irregular mass of ore, apparently filling cavernous spaces in the earth's crust, and showing no indications of being connected to other mineral deposits. From its entire want of any order, its deposition would seem to have been accidental, rather than the result of any well-recognized system.

Veins of this class are sometimes found in stratified rocks, between the strata, and sometimes enveloped in unstratified rocks.

REMARKS.—Besides the several classes of veins described, small masses of rock, termed " nests," " concretions," " nodules," bearing metallic substances, occur in the middle of strata, or beds of other rock. The ores of iron are frequently found in these forms.

False veins also occur in addition to the veins enumerated. These, as to their " walls " and " dip," resemble proper veins; but being composed of sand, clay and other alluvial substances, are readily distinguished from the true crystalline and mineral veins.

FORMATION OF MINERAL VEINS.

The evidence is abundant that many mineral veins were a long time in being formed after the fissures in which they exist were produced.

Banded veins, composed of crystalline matter in the form of plates or " combs," divided from each other by thin layers of clay, conclusively prove, not only that their formation was slow, but that several re-openings of the same fissures must have taken place at periods separated by long intervals.

The fissures occupied by metallic veins seem to correspond to such rents as are produced from time to time by the shock of an earthquake, and in a majority of cases are evidently attributable to the same causes.

It is not improbable that some of these fissures were produced by contraction, while the rocks through which they run were passing from the plastic to the solid state. The manner in which they were filled with ore and vein-stone is still a subject of speculation among geologists. The principal theories advanced to explain the phenomena of mineral veins are as follows :

1st. That open fissures in the earth's crust were filled with crystalline and metallic matter by aqueous infiltrations from above.

Thus, Werner supposed the substances of mineral veins to have been precipitated from a " *chaotic menstruum*" into fissures in the earth while it was in some nascent state. This theory at present has less advocates than formerly.

However, that false veins were produced by infiltrations from the surface of the earth is evident from the character of the deposits. It is also evident that some of the material found deep in proper veins came from the earth's surface or from the bed of the sea. Thus, for example, well-rounded pebbles of quartz and slate, pieces of coral and marine fossil shells occur imbedded in several mineral veins at depths in some cases of over a thousand feet below the surface. But these occurrences without doubt were accidental, and had no connection with the general system of repleting proper veins.

2d. That the contents of mineral veins, like those of dikes, were formerly molten, and while in this state were injected from below by mechanical force.

This view of the subject was proposed and maintained by Dr. Hutton.

The explanation which it affords relative to the productions of some mineral veins is quite satisfactory. But the evidence seems insufficient to warrant the inference that any considerable proportion of metallic veins was produced in this manner.

3d. That the contents of some mineral veins were deposited in fissures and cavities in the earth's crust by the condensation of mineral vapors emanating from vast subterranean fountains of intensely heated matter. This theory, suggested by M. Neckar and Dr. Buckland, appears to be an improvement upon that of Dr. Hutton's, especially in its application to the formation of veins of banded structure. The repletion of veins of this class, owing to the great length of time requisite for its accomplishment, must be referred to chemical agencies rather than to mechanical forces.

4th. That some mineral veins were produced in the same manner as flint, or other concretions, by chemical segregation from the inclosing rock prior to its solidification. It not unfrequently occurs that veins are found passing by insensible gradations into the containing rock. As this could take place only while both were in the liquid state, it is evident that such veins are of cotemporaneous origin with the rock in which they are inclosed, and that they were in the process of more complete segegration at the time of their becoming solid. This theory, proposed by Prof. Sedgwick, furnishes a solution quite satisfactory as to the method by which many of the flat or dilated veins, also the accumulated veins and other masses of isolated mineral substances were produced. And it is not improbable but that the principle of chemical segregation was active in the production of some proper veins.

5th. That the formation of many mineral veins is due to electro-chemical agencies.

The existence at present of electro-currents in some Cornish veins, and the analogy between voltaic combinations and the arrangement of matter in mineral veins are adduced by R. W. Fox and M. Becquerel in support of this theory. Mr. Fox goes so far as to infer that the richer veins in metallic treasure are at right angles to the earth's magnetism.

It is more than probable that electro-chemical influences are capable of transforming to considerable distances the substances composing mineral veins, even from the solid rocks in which they are disseminated into fissures in the same, and of determining the arrangement of the matter so transferred; but that electro-magnetic currents effect richer metallic deposits in veins running east and west than in those running north and south is refuted by unequivocal evidence. True, the principal and richer veins of Cornwall run east and west, but the principal veins of the entire Pacific Slope run nearly north and south.

6th. That the matter of mineral veins originally disseminated in portions of the rocks adjoining rents and cavities in the earth's crust, was introduced into these openings by the process of infiltration. That the filtered matter may have been derived by simple solution from the inclosing rocks, or by decomposition of some of their constituents; and when gathered into the receptacles may have concreted unaltered in composition, or have undergone various changes, induced by the mutual action of its component parts. It has been inferred that most mineral veins were found in those fissures formerly occupied by thermal springs, for much of their material is identical in character with that composing the walls of those springs, or held in solution by their waters. Besides, the vast depths to which both mineral veins and thermal springs extend downward, and the similar relations which they respectively bear to the great lines of upheaval and dislocation of rocks, go far in support of this inference. That some mineral veins are of recent formation, by infiltration, and that others are in process of formation by hydro-thermal agencies, seems no longer problematic.

Geological researches on the Pacific coast disclose in the vicinity of the village of Volcano, Amador county, California, "distinctly marked quartz veins cutting through the gravel, and evidently formed by the action of water holding silica in solution;" also, that in Steamboat Valley, Nevada, semi-crystalline depositions of silica and oxide of iron are being made on the walls of the fissures of the boiling springs.

In this vicinity, another fissure of the same origin is still nearer replete with mineral matter. This fissure, at a few points along its extent, evolves steam and carbonic acid, but no water. For the most part it is filled with silicious concretions which contain the oxides of iron and manganese, the sulphides of iron and copper, and metallic gold.

EXPLORATION OF MINERAL VEINS.

Many rules have been made for the *Exploration of Mineral Veins*. But these, so numerous are the exceptions, are to be regarded hints rather than rules. Mineral veins occur in nearly every other formation of rocks; and no unexceptional law seems to prevail as to their comparative richness in the different formations. Not unfrequently a mineral vein runs horizontally through different strata of rock. Richer metallic deposits are sometimes found in the vein, inclosed in a stratum of one formation of rock and sometimes in a stratum of a different formation. Several authors are of opinion that metallic veins are richer near the junction of stratified and unstratified rocks. In several quartz-mining districts of California—to wit, in Mariposa—some of the gold-bearing quartz veins are in granite, but the larger, and so far as determined the richer veins, are in slate; in Tuolumne the more valuable veins occur in granite, those in the metamorphic slates are comparatively barren; in Calaveras the quartz veins are found valuable both in granite and limestone; in the Alta mining district the "wall-rock" is serpentine, white talc accompanies the richer deposits; in Amador the valuable veins are mostly in slate. The celebrated Eureka, Oneida, Keystone and Spring Hill veins are in slate. The former has been worked nearly thirteen hundred feet in depth. The vein, six feet thick at its "outcrop," has regularly increased in thickness and richness with the sinking.

The following suggestions are esteemed among the first of importance for the *Exploration of Mineral Veins:*

1st. That mineral veins are more abundant in the older crystalline rocks than in those of more recent formation.

2d. That they are more abundant in mountainous and hilly regions than in the even land.

3d. That they are most productive near the junction of stratified and unstratified rocks.

4th. That their locality is often indicated by the color of the surface of the land. The color arises from the decomposition of some of their constituents.

5th. That their locality is frequently determined by fragments of the vein scattered over the surface of the ground.

6th. That they are discoverable by their "outcrops." The "outcrop" is the most obvious indication of a mineral vein. It is doubtless true that the great majority of mineral or metallic veins discovered, is due to accident rather than to science, and that the discoveries so made, to a great extent, have been by means of the "outcrop." Sometimes the vein-stone and sometimes the "walls" or country-rock, is the more obdurate. As the case may be, the more yielding will have been worn away by the action of water and other agencies, leaving the other, so that the exploration of the vein is rendered quite simple.

7th. That when the bearing is approximately determined, the vein may often be found by sinking one or more shafts through the alluvial deposits into the underlying rock, and making a "trench" or "tunnel" at right-angles to the bearing of the vein.

EXPLOITATION OF MINERAL VEINS.

The *Exploitation of Mineral Veins* comprises the various processes by which their ores are won from their natural position and brought to the surface of the earth.

The Workings are Open and Subterraneous.

Open Workings apply to flat or dilated veins, and other horizontal deposits situated near the surface. They are usually arranged in terraces suited to the character and disposition required of the material. Open workings are sometimes quite extensively employed in the devleopment of mineral veins of considerable " dip." That is, the vein, in the direction of its bearing, is uncovered for a long distance. This practice, however, in most cases, cannot be too severely reprehended. The natural roofing of the vein should not be interfered with further than absolute necessity requires.

Subterranean Workings are of two kinds, the *Preparatory* and those of *Extraction.*

The *Preparatory Workings* consist in timbered or otherwise properly secured shafts and tunnels, exposing most effectually the ore sought to the attacks of the miner; also in the means for ventilation, for the discharge of water, and for the transportation of the detached mineral out of the mine.

A *shaft* is a pit or narrow opening into a mine, and may be vertical or inclined. It is to be regarded a tunnel when it forms with the horizon an angle less than forty-five degrees (45°).

A *winze* is a small shaft connecting two levels, and is employed for the purposes of ventilation or of explora-
·tion.

The *working-shaft*, termed also the "hoisting-shaft," is employed for lowering into the mine, timbers, "spiling" supplies and tools, also for raising ore, water, etc. The miners, for the most part, pass in and out of the mine through this shaft—either by ladder or in some form of carriage, driven by steam or other motive power. The working shaft is seldom less, transversely, than thirty by forty inches. Ordinarily it is much larger—of sufficient size to admit the use of the double hoisting track, pumping machinery, and miner's ladder.

The *air-shaft* is employed for the purposes of ventilation; also not unfrequently for filling with dirt and debris the chambers from which the ore has been extracted. The working and air shafts are occasionally several hundred feet apart—not often less than one hundred feet.

A *sump* is a pit usually at the bottom of the working shaft, and is several feet deeper than the lowest of the other workings. Its chief use is that of a reservoir into which the waters of the mine are drained, whence by machinery they are discharged at the surface.

A *tunnel* is a narrow opening into a mine, and may be horizontal or inclined. It is to be regarded a shaft when it forms with the horizon an angle greater than forty-five degrees (45°). A tunnel excavated in a vein, and running the vein's course, in some localities is called a "gallery." A tunnel piercing or intersecting a vein is termed an "adit." A tunnel connecting two shafts is called a "drift," or "drift-way." Tunnels are seldom less than two feet by three and one-third feet high. Quite a common size is three feet wide at the top, three and a half to four feet wide at the bottom, and six feet high. The size of the tunnel is very frequently determined by that of the vein in which it is driven—the width of the former being made equal to the thickness of the latter.

In considering further the problem of *Preparatory Workings of Mineral Veins*, let the rake vein be taken, since, on account of its greater extent and regularity, it is subject to more general and systematic operations than any of the other forms. Regarding the occurrence of the vein in even land, two shafts—working and air shafts—should be sunk to nearly the same level, and united at their greatest depth by a tunnel. The sump, the mine containing much water, should be sunk before the connecting tunnel is driven far—the shafts should be at least a hundred feet apart, and, independent of the sump, fully a hundred feet deep, and should, with the tunnel, be well timbered or otherwise properly secured.

The Preparatory Workings may be almost indefinitely extended by sinking the working and air shafts, and connecting them by tunnels at desired intervals; also by sinking on the line of the vein other shafts and connecting them by tunnels. The "sinkings" or "distances" between tunnels are usually from sixty to one hundred feet.

Regarding the occurrence of the vein in a mountain, tunnels may often first be employed to advantage. The vein and its inclosing mountain ridge being parallel and

unbroken, an "adit" at the lowest feasible level should
be made to intersect the vein, thence a gallery driven on
the vein to some desired point from which a shaft should
be raised or excavated to the surface. In case of a break
occurring, as by a cañon, the work frequently is greatly
facilitated, as thereby a tunnel may be commenced upon
the transverse or broken face of the vein and driven
directly on its course. This use of tunnels should rarely
be adopted, unless the elevation exceeds a hundred feet
in height.

As a general rule, shafts and tunnels should be wrought
in the "lode."

Veins, however, very thick and friable, having a dip
of less than forty-five degrees and being inclosed by unstable walls, furnish exceptions to this rule; such are to
be worked by vertical shafts rather than by shafts coin-
ciding with the dip of the vein.

Workings of Extraction consist in detaching the vein-
matter from its natural position and hoisting or bringing it to the surface of the earth.

There are two modes of attacking the vein, viz.: by
Direct or Descending Steps, and by *Reverse or Ascending
Steps.*

In Fig. 1, taken in the plane of the "dip" and "strike"
of a mineral vein of the Rake species, let A B represent the working-shaft, C D and E F air-shafts, B D
and B F tunnels or galleries, G H and I K the limits of
the workings toward the surface, S the sump. Con-
ceive the vein to be laid out into parallelopipeds or rec-

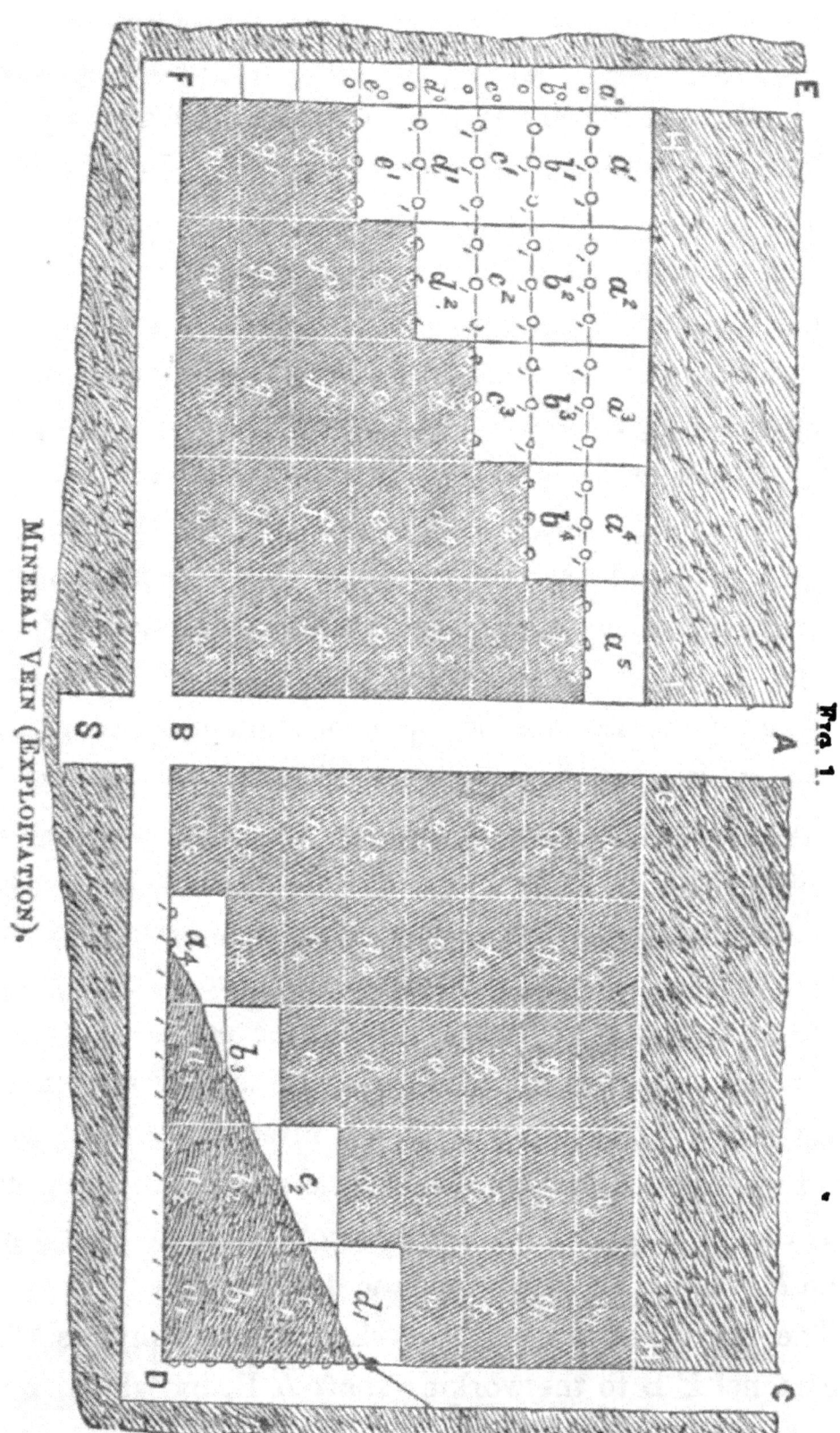

FIG. 1.

MINERAL VEIN (EXPLOITATION).

tangular masses, and let these masses be represented by $a_1 - a_5$, etc., to $n_1 - n_5$; $a^1 - a^5$, etc., to $n^1 - n^5$. One of these masses is commonly fifteen or twenty feet long, six or eight feet high, and as wide as the vein. Let a^o, b^o, c^o, d^o, e^o, represent scaffolds or stages, o_1, o_1, o_1, floors firmly timbered employed as tracks or ways for conveying the ores to a common point of delivery, also for sustaining the rubbish and securing the walls of the mine—and let x_1 represent a guide-board or " shute."

By Direct or Descending Steps.—To extract the ore by " descending steps," usually two miners, " leader " or " lead-hand," and " striker," termed " a set," commencing on stage a^o, excavate in succession the parallelopipeds or masses a^1, a^2, a^3, a^4 and a^5, deliver them into the shaft E F, and construct the floor o_1, o_1, o_1, shown at the bottom of the excavated chambers $a^1 - a^5$. In like manner a second set, commencing on stage b^o, excavate and dispose of the masses b^1, b^2, b^3, b^4 and b^5; and so on set after set of miners are employed, and the several masses from a to n are disposed of and the respective floors o_1, o_1, o_1, constructed as required. The order observed is that each lower set of miners succeeds the set next above by the length of a parallelopiped, that is fifteen or twenty feet. Thus when five sets have come to be employed, while the upper set is engaged on the mass a^5, the next lower set is on b^4, the next lower on c^3, and so on to e^1.

The ore delivered into the shaft E F is conveyed by the tunnel F B to the working-shaft A B, by which it is raised to the surface. It may be remarked that the shaft E F is sometimes employed as a hoisting-shaft, also that the excavations are sometimes made without the use of the floors o_1, o_1, o_1; such workings, however, are crude and reprehensible.

By Reverse or Ascending Steps.—To extract the ore by "ascending steps," a set of miners, commencing on a temporary stage, excavate in succession the parallelopipeds a_1, a_2, a_3, a_4, etc., and firmly timber floor or roof as shown by o_1, o_1, o_1, the tunnel B D. In like manner a second set excavate successively the masses b_1, b_2, b_3, etc.; and so on set above set of miners, from a to n, are employed.

The order observed is that each upper set of miners succeed the set next below by a parallelopiped fifteen or twenty feet in length. Thus when five sets have come to be employed, while the lower set is engaged on the mass a_5, the next upper set is on b_4, the next upper on c_3, and so on to e_1. Temporary stages are chiefly used.

By this method of reversed steps, the ore, as detached, falls by its own weight. It is sometimes allowed to fall freely into the tunnel, but is usually retained in the mine, resting upon the tunnel floor or roof, whence it is discharged as required into cars, conveyed to the shaft A B, and thence hoisted to the surface.

It may be observed that the shaft C D, or more generally speaking the shaft from which the "stopings" or excavations are carried on, is sometimes employed as a hoisting-shaft. This resort, however, should only be in extreme cases.

The walls of a mineral vein, especially if excavated by ascending steps, can often be more effectually and economically secured by filling the chambers with dirt and debris, obtained at the surface of the earth, than by any other means.

When the inclosures are very friable and unstable, the process of filling should succeed immediately that of extraction. The miner thereby is not only secure against the crumbling and falling walls, but is brought directly up to his work.

The filling material, for example, is thrown into the air-shaft CD, falls upon the adjustable shute x, whence it is conducted into the openings $a_1 - a_4$, $b_1 - b_3$, $c_1 - c_3$, etc. The filling should be well tamped.

In considering the relative merits of the two modes of working, it is to be remarked:

That the mode of "descending steps" affords the miner less restraint in position, and greater freedom of action; that it is better adapted to the sorting of ores in the vein; and that it is, in case of the vein-stone being much decomposed, less subject to loss by the liberated minerals or metals becoming mixed with the rubbish.

And on the other hand: That the mode by "ascending steps" is more expeditious and cheaper, requiring less expense in timbering, flooring and filling, and less labor in excavating, since the ore, by its own weight, is more readily disengaged from its natural position.

Again, as one of the chief objects in mining should be to keep the mine in good order, both regarding the facility in operations and safety to the workmen, the caution cannot be too well heeded against breaking through the surface, or weakening the natural roof of the vein. In most cases the "limit of workings," G H, I K, depending on the nature of the ground and thickness of the vein, should not approach the surface nearer than at a distance of twenty feet. Pillars of vein-stone are sometimes left in the mine for its better protection. These may be left, especially along the sides of the working-shaft, with the highest advantage. They, together with the upper portion of the vein, can readily be reclaimed, whenever desired.

As a general rule, it seems proper to observe, that on engaging in quartz mining, the first thing to be done is to thoroughly "prospect" the vein or mines. To accomplish this, regarding the vein of the Rake species, a shaft should be sunk on it not less than one hundred feet deep, a tunnel driven as far on its course, and the ore worked by a process near as possible to that proposed to be adopted. If the results prove favorable the adventurer may proceed to erect his quartz machinery for working on a large scale, with assurances of almost certain success. If the results are unfavorable he may congratulate himself that he has not fallen into the too frequent error of building expensive mills on worthless "lodes."

VENTILATION OF MINES.

The Ventilating of Mines is the act of circulating through any or all portions of their " workings" pure air, and substituting it for foul air and other noxious substances, of which those of most common occurrence are " choke-damp" (carbonic acid) and " fire-damp" (carburetted hydrogen) each producing by inhalation almost instant death. The former of specific gravity 1,529 occupies the lowest part of the mine's workings, and the latter of specific gravity ,555 the upper part. " Fire-damp" is combustible, and a mixture of one volume of it with ten volumes of air is highly explosive. It is the burning of this gas, and the explosion of its mixture with air, that occasion at times such fearful havoc of life and property. The resulting carbonic acid is called " after-damp," and is more fatal to miners than even the fire and explosion.

Choke-damp and Fire-damp do not usually occur in any considerable quantities in the same mine. They are especially abundant in deposits of coal, which, being broken, one or the other is often copiously evolved from its immense store-house. Their occurrence in mineral or metalliferous veins in some cases may be referable, to no little extent, to the decomposition of vein matter, but for the most part is attributable rather to the respiration of the workmen, the combustion of candles and blasting powder, and the decomposition of wood and filling.

On entering a mine in which the existence of "choke-damp" is suspected, the adventurer should carry, on the end of a pole, a burning candle, thrust a few feet in advance. If the candle emits a strong light he can safely proceed—if a feeble light, his progress is attended with peril—and if it ceases to burn he cannot too quickly retrace his steps, for certain death awaits him.

The existence of "fire-damp" in a mine is indicated by the behavior of the "top" and flame of a candle, which should be well trimmed.

The "top," or haze-like cap of the flame, maintaining a yellowish brown color, not exceeding a quarter or half inch in length, denotes that the air is almost or quite free from "fire-damp." The "top" elongating and changing to a bluish-gray color warns the miner of the presence of this gas. The "top," as also the flame shooting up into a sharp spire, the former changing its color from bluish-gray to a clear blue, while minute, luminous points rise rapidly up through the flame and top, indicate that the place abounds in "fire-damp," just on the point of bursting into flame and explosion. The miner in this perilous condition, mindful that the least agitation may effect the ignition of the gas, slowly lowers his candle to the floor, and cautiously retreats or ventures to extinguish the flame with his thumb and finger.

The means of the Ventilation of Mines are "natural" and "artificial."

Natural Means. The difference between the temperature of the mine and that of the external air, combined with the difference between the levels of the openings at the surface, also at the points of their connection below ground, produce a circulation of air through the mine.

A mine having two shafts of different levels at the surface, and of the same level at their points of tunnel-connection at the bottom, in case the internal temperature is greater than the external, a circulation will be established through the mine. The rarified air will rise and pass off, as by a chimney, through the shaft of the more elevated opening, while its place will be supplied with fresh air flowing in through the shaft of less elevated opening. But in case the external temperature is greater than the internal, the direction of the current will be reversed.

If the mine occurs in a level country, a high chimney, reared over one of the shafts, is made to supply the lack of a natural difference in elevation.

Experience, however, shows that a mine having two shafts of the *same level at the surface*, and of different levels at the points of their tunnel-connection below, a current of air will flow quite uniformly down the deeper shaft, along the tunnel, and up the shaft of less depth.

A long tunnel driven into the side of a mountain may readily be ventilated, by dividing it into two unequal parts by a horizontal partition, by connecting its upper and smaller part at some distance from the entrance, with a shaft raised to the surface, and by letting its lower part communicate with the entrance and with the upper part at the other extremity of the tunnel. A current is produced as the air in the upper or smaller part becomes sooner heated by the enclosing rock, and passes off through the shaft, while the fresh air to supply its place flows in at the entrance, and thence along the lower part of the tunnel to its extremity.

A mine may be ventilated to a limited extent by the application of the wind. The apparatus consists of a turn-cup, having a funnel-mouth open to the windward, and communicating with the mine through a pipe. The wind, arrested by the machine, is conducted by the pipe into the mine, and disposed of as required.

Artificial Means. In artificial means are involved the principles of " propulsion and exhaustion." In applying the principle of " propulsion," large bellows, force-fans, and sometimes force-pumps are employed. By these, the air, through pipes or other conductors, is driven into the mine to those parts where most needed, whence it is dispersed through the workings, and finally escapes at the mouth of the opening. This principle is especially advantageous in that it readily admits the use of flexible pipe, or hose, by which the air is conducted with facility into recesses of most difficult access. Its application, however, is either on a small scale or only comparatively temporary. The principle of " exhaustion " is more generally applied, and with far greater economy and effect than that of " propulsion." It is put in practice by " exhausters," " fires " and " furnaces." These, communicating

2*

with the open air, are in some way connected with pipes, or their equivalents, extending to the remotest parts of the mine.   As the machines are put in operation a circulation of air is produced through the mine, the current extending from the "opening" to the extremities of the exhaust pipes, through which it then, fraught with impurities passes and escapes above ground.

Instead of ordinary pipes, there are sometimes substituted covered trenches dug in the wall-rock of the shaft and along the bottom of the tunnel.   The rock furnishes three sides of an air conductor.   The fourth is formed of plank fitted in its place air-tight with cement, or clay. Sometimes the shafts and tunnels entrie are converted, as it were, into air-pipes.   This is especially the case in the more extensive works.

The "exhausters," so called, consist of various contrivances, for the most part in the form of "pumps" and "fans," whose relative merits with one another and among themselves it is not proposed to discuss.   "Fires," for the purposes of exhaustion, are applied in different ways—by being placed at the mouth of the "up-cast" or chimney-like shaft, or at the bottom, or by being suspended in grates part way down the shaft.

"Furnaces," especially in large operations, are most in use.   The position of the furnace is in connection with the "up-cast" shaft, whose office is to conduct away the vapors, vitiated air and gases of the mine, as they are put in motion and caused to rise by the exhaustive influences of the heat.   The furnace is sometimes placed at the top and sometimes at the bottom of the shaft. The plan of placing it at the bottom is preferable, since, in comparison, the sides of the shaft becoming heated, ventilation is continued longer, and is less liable to derangement, in case of accident occuring to the furnace. The "up-cast" and "down-cast," in other words the escape and supply air-shafts are, in effect, the one at one end of a long tunnel, and the other at the other end. Other than this they have no communication, though frequently only separated by a partition wall. In connection with the furnace it is now proposed to illustrate a system of ventilation which is applied with complete success even in coal mines abounding in fire-damp or other noxious substances.

Let the mine be developed by a system of rectangular tunnels, represented one way by letters *a, b, c, d, e,* and the transverse way by numbers 1, 2, 3, 4, 5, 6, 7, and by two shafts termed the "up-cast" and "down-cast," sunk at the junction of *b* and 4 tunnels. Let air-tight doors be placed in the tunnels as follows :—in *a* between 5th and 6th; in *b* between 1st and 2d, 2d and 3d, 6th and 7th; in *c* between 2d and 3d, 3d and 4th, 4th and 5th; in *d* between 2d and 3d, 3d and 4th, 4th and 5th; in *e* between 3d and 4th ; in 2d between *c* and *d*; in 3d between *a* and *b*; in 5th between *b* and *c, d* and *e*; in 6th between *b* and *c, c* and *d, d* and *e,* and in 7th between *c* and *d.* Let the "supply" opening in the "down-cast" shaft be in 4th looking toward *a*; and let the *escape* openings in the "up-cast" shaft be—one in *b* looking toward 3d and one in 4th looking toward *c*. On the furnace being put in operation the fresh air descends the " down-cast " shaft, issues through the supply opening, and passes down tunnel 4th to the junction of 4th and *a*, where it divides into two streams. One stream flows to the right along *a* up 5th, along *b* down 6th, along *a* up 7th, along *c* to the left up 5th, along *d* to the right up 7th, along *e* to the left down 4th into the " up-cast," and effects its escape. The other stream flows to the left along *a* to 2d, then in two branches—one up 2d and along *c* to 1st, the other along *a* and up 1st to *c*; the branches here uniting, the stream flows up 1st to *d*, then in two branches —one along *d* and up 2d to *e*; the other up 1st and along *e* to 2d;—the branches here uniting, the stream flows along *e* down 3d and along *b* into the " up-cast," and escapes by the chimney.

In case of "fire-damp" being very abundant in any portion of the mine, it is conducted by the way of the dumb-furnace, which communicates with the " up-cast " shaft, out of reach of the flame. The draught of this furnace arises from the exhausted condition of the *flue* of the active furnace.

Too much care, however, cannot be used to dilute with fresh air this gas below the explosive point before suffering it to escape. Explosion does not occur when the volume of air is less than six or more than fourteen times the volume of the gas. "Fire-damp" should not be inhaled unless mixed with air of full thirty times its volume. In practice, the mixture becomes not less diluted before it is taken to the active furnace. The doors or valves are double, that is, placed at some distance apart, so that by the alternate opening and shutting of them the currents of air are not interfered with.

The "doors," as also furnaces of ample capacity being properly arranged and managed, the ventilation of the mine, in its every part, is brought completely under control. At the pleasure of the manager, vapors the most pestilential, and gases the most poisonous and inflammable are aerified to a degree of entire salubrity, or imprisoned for the time being, or sent harmless away by themselves. This system applies alike to coal and horizontal deposits, or to mineral veins of any dip and extent,—in a word, is capable of almost unlimited application.

QUARTZ MACHINERY.

For the mechanical reduction and concentration of ores, two kinds of quartz machinery, the *wet* and the *dry*, are in use. The application of the former obtains in nearly or quite all extensive works, while that of the latter is mostly limited to the test-room and experimental operations.

For the full treatment of ores, auxiliary machines, common to the two, are employed. Of these the Rock-Breaker, Amalgamator and Separator may be mentioned. The most efficient and economical quartz machinery thus far introduced for the mechanical reduction and concentration of ores and rock containing the precious metals, and for the purposes of amalgamation and subsequent treatment, consists, aside from prime movers, common connections and tanks, chiefly of the Rock-Breaker, Straight Batteries, Grinders and Amalgamators, Separators, Concentrators, Retorts, Crucibles, and Ingot Molds.

ROCK BREAKERS.

Rock Breakers are machines chiefly employed for the crushing of rock, as it comes disengaged from the mine, quarry or other bed, to fragments an inch, half inch or so in diameter. For this work they are fast superseding the "Cornish Rollers" and "Heavy Stamp"—in fine, all other machinery designed for like purposes. They have also been employed for the pulverization of ores and rock, but have not been sufficiently successful to warrant their use in this department.

The nearly upright jaws with convergent crushing faces constitute one of the chief features of Rock Breakers. These are caused by means of machinery crank, or its equivalent, directly or indirectly applied, alternately to recede from and approach each other. In some machines one, and in others both jaws are movable; and in some cases they have, additional to the vibratory, a lateral or vertical grinding or other compound motion. The "Blake Stone Breaker" has one of its jaws stationary, the other movable. In the upper end of the latter is the centre of motion.

The inventor, in describing the operation of the machine, says " every revolution of the crank causes the lower end of the movable jaw to advance toward the fixed jaw about $\frac{1}{4}$ of an inch and return. Hence, if a stone be dropped in between the convergent faces of the jaws, it will be broken by the next succeeding bite; the resulting fragments will then fall lower down and be broken again, and so on until they pass out at the bottom. The ditsance between the jaws at the bottom, which limits the size of the fragments, may be regulated at pleasure."

The " Blake Stone Breaker " may be mentioned without any disparagement to others of its class as having made for itself, in the Pacific gold and silver regions, an enviable reputation. One of these machines, with a feed opening of nine by fifteen inches area, is found capable of reducing from one hundred to one hundred and fifty tons of hard rock per day to a size suitable to be fed into Batteries.

The machine of the size mentioned weighs 11,600 pounds, and requires six-horse power to reduce seven and one-half tons of rock per hour.

BATTERIES.

The *Battery* embraces the frame (usually made of wood), Mortars, Cams, Cam-Shaft, Tappets, Stamps, Stamp-Stems, Shoes, Dies, Guides and Screens.

MORTARS.

High Mortar.—The High Mortar is represented by Figs. 2 and 3. Fig. 2 shows a front and Fig. 3 an end view of the mortar mostly approved on the Pacific coast and wherever else it has been employed for the reduction of ores containing the precious metals. B represents the feed opening, and 'C the screen opening. The screen is fastened by nails or screws to the screen-frame, which by wedges is firmly secured in grooves provided for its reception at the ends of the mortar, and by two " lugs," seen at the bottom of the opening C.

Mortars, for working on a large scale, are constructed

for three, four, five or six stamps each. Millmen and quartz operators, from long experience, are unanimous in opinion that the five-stamp mortar is preferable.

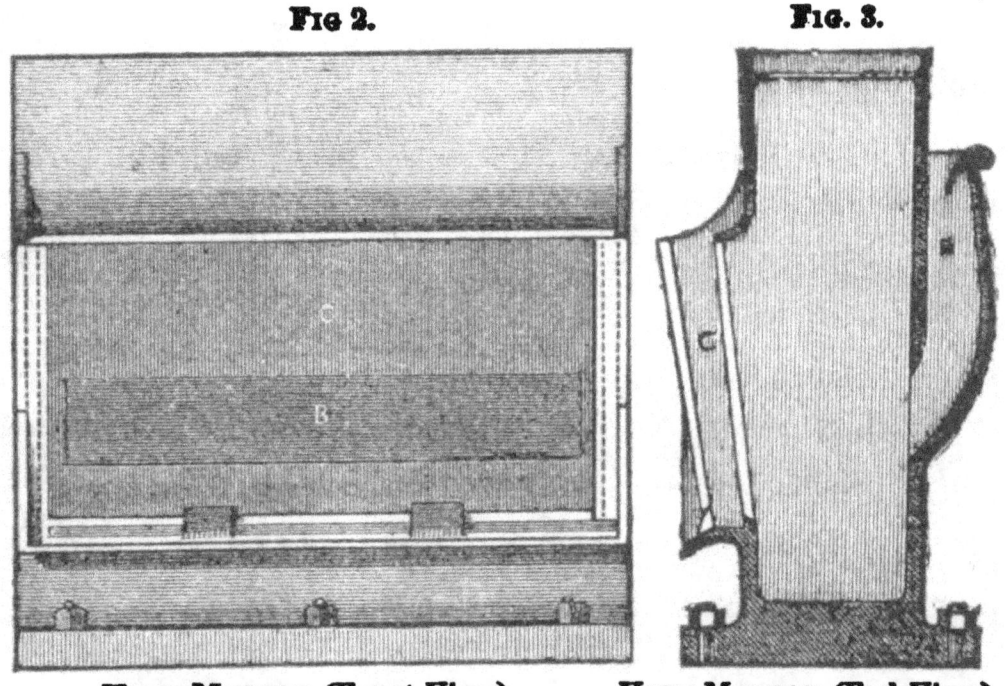

HIGH MORTAR (Front View).   HIGH MORTAR (End View).

Section Mortar.—The Section Mortar, designed for places of difficult access, can readily be taken apart and packed on mules. The Section Mortar may be high or low. It should only be employed in extreme cases, owing to its cost and greater liability to destruction than that of the Solid Mortar.

Double Discharge Mortar.—Fig. 4 shows a side and Fig. 5 an end view of the Double Discharge Mortar. A A represents the doors, B the feed opening; C the screen. The doors A A are lined on the inside with copper plate, as shown by the thick black lines. This style of mortar is designed more especially for the working of gold bearing rock. By removing the screen C and carrying the door somewhat higher, we have what is known as the Float Mortar. By substituting the door A for a screen, we have the High Mortar, as shown in Figs. 2 and 3, which is in most general use for the working of both gold and silver ores.

FIG. 4.   FIG. 5.

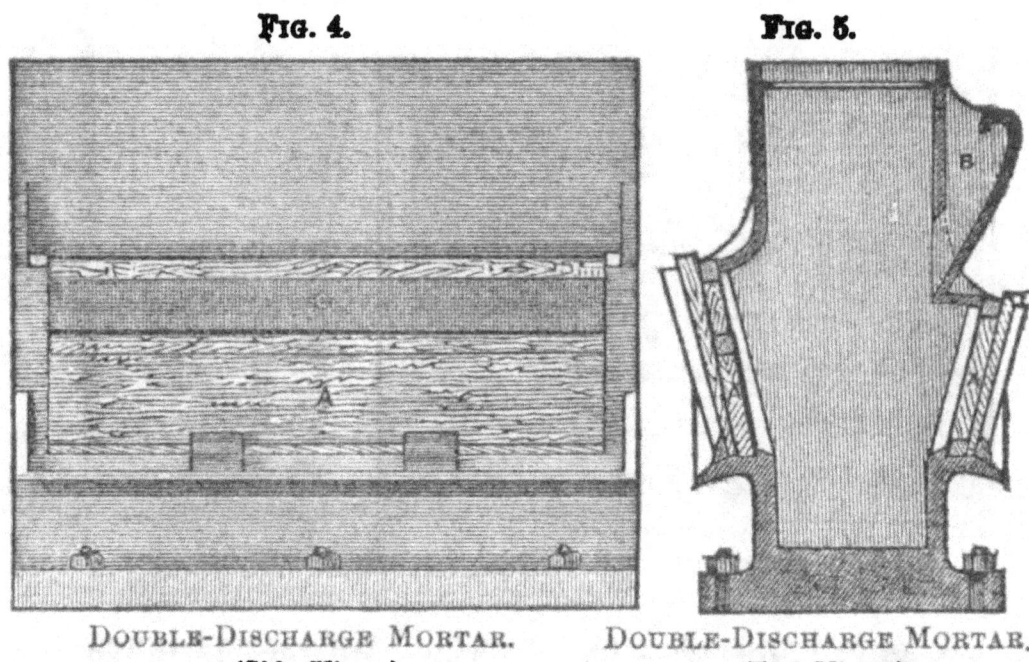

DOUBLE-DISCHARGE MORTAR.   DOUBLE-DISCHARGE MORTAR.
(Side View.)   (End View.)

Dry Mortar.—The chief difference in construction, usually, between the "wet" and "dry working mortars" is that the screens to the latter are higher or wider and at a less inclination to the horizon than in the former, besides being made of wire cloth instead of Russia iron. In the "Wheeler and Hotaling Dry Mortar" no screen is used; in this machine the discharge-opening is about six inches wide and between two and three feet above the die. The ore pulverized very fine, rises to the discharge-opening, and thence passes through an *exhaust fan* or *blower* into the settling chamber, from which an escape or air-supply pipe leads back to the battery. It has been proposed to employ a current of steam instead of the *fan*, but not aware of its having been put in practice, the writer cannot confidently speak of its worth.

The Low Mortar.—The Low Mortar is less used than formerly. Its use, however, is far from being abandoned. The most improved low mortar is constructed with frame-grooves and discharge-lips.

TAPPETS.

The Tappet is the head, or machine secured to the stamp-stem, by which the cam raises and lets fall the stamp upon the substance to be crushed. Fig. 6 represents the Plan, and Fig. 7 an elevation of the "Gib-tap-

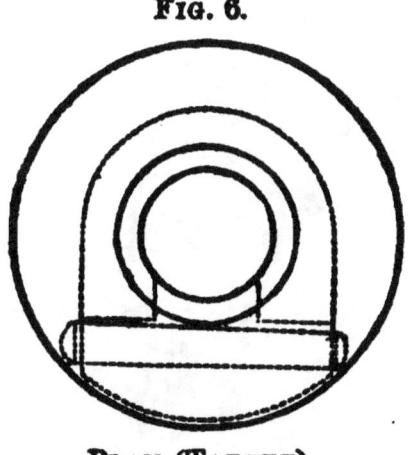

FIG. 6.

PLAN (TAPPET).

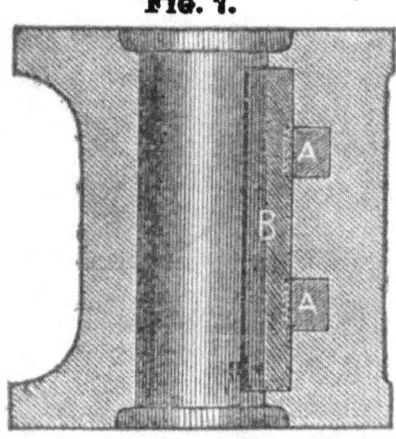

FIG. 7.

ELEVATION (TAPPET).

pet "—one of the many very valuable inventions in quartz machinery of Zenas Wheeler, Esq., of Wheeler & Randall. B, in Fig. 7, represents the "gib," and A A the keys by which the tappet is firmly secured to the stem.

By this invention, no cross and weakening key-seats in the stem are required; besides, it can readily be adjusted to any desired point. It surpasses in excellence any other in use for like purposes.

CAMS.

The Cam is a machine employed to raise the stamp to any desired height, and then allow it to fall by its own gravity upon the substance to be crushed. Cams, in quartz machinery, are constructed with one, two or three arms. The "three-armed" cam is but little used, as by it

FIG. 8.

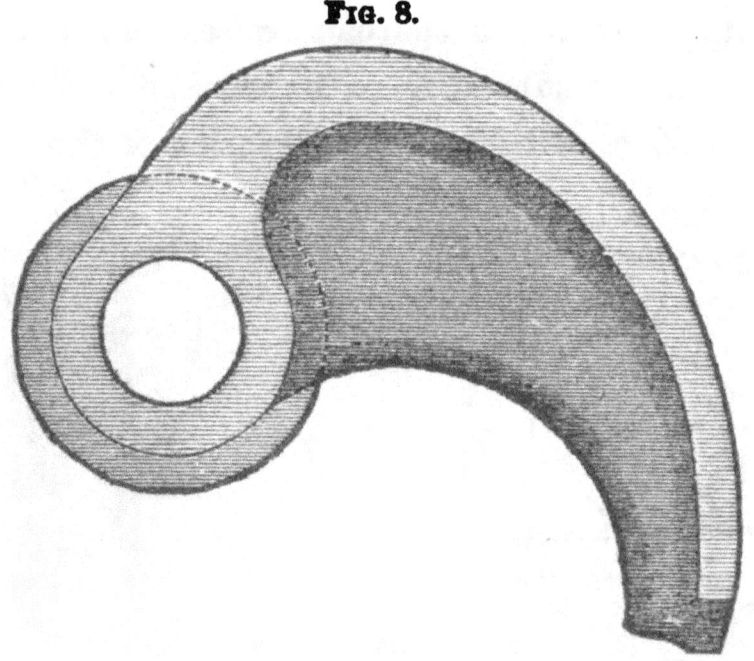

SINGLE-ARMED CAM.

a less number of drops to the stamp can be obtained than by either of the other styles mentioned. Fig. 8 represents the "one" or "single-armed" cam, and Fig. 9 the "two" or "double-armed" cam. By the "single-armed " cam, when properly constructed, ninety ten-inch drops per minute are obtained with safety and economy.

The proper curve of the cam-arm, so as to run with the least friction, is a modified involute of a circle, whose radius is equal to the horizontal distance between the centre of the cam-shaft and the centre of the stamp-stem. The cam-arm should have a greater or sharper curvature near each of its ends than the regular involute of the circle. The objects of this are: 1st, to receive the tappet at the least practical distance from the centre of the cam-shaft, where the concussion is less than at a greater distance; and 2d, to prevent the sharp end of the cam from tearing along the face of the tappet, from a point near its centre to its outer edge

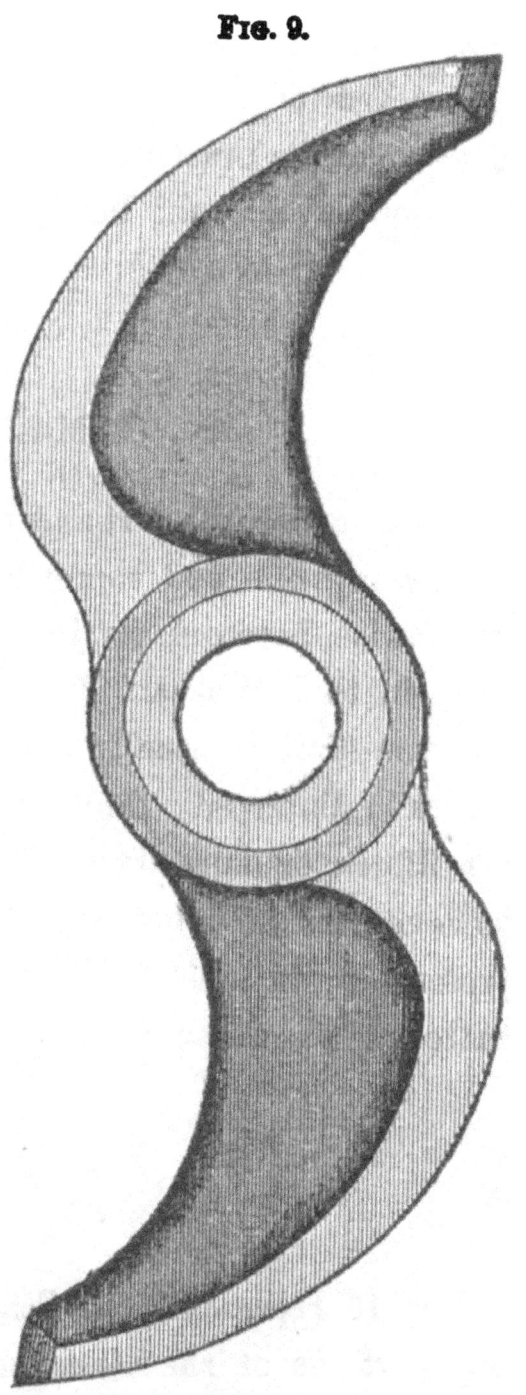

Fig. 9.

DOUBLE-ARMED CAM.

or point of delivery. The outer end of the cam-arm is

fashioned so as to conform to the outer edge of the tappet, which is circular.

It seems proper to remark, since the curve of the cam-arm is determined in reference to the distance between the centres of the cam-shaft and stamp-stem, that this distance in the practical erection of the machinery should never on any account be changed in respect to the same curve.

CAM-SHAFT.

The Cam-Shaft is a round bar of iron, turned, finished, and having one and sometimes two key-seats cut in it lengthwise, between its journals, for the securing of the cams in their places. For stamps of ordinary weight, the cam-shaft is usually made about four and one-half inches in diameter. Sometimes one shaft is employed to run fifteen or more stamps. An independent cam-shaft, however, for each battery of five stamps is much more preferable; in which event, if there is a line of several batteries, a counter-shaft is usually employed.

STAMPS.

Fig. 10 represents the Stamp-head or socket, including portions of the stamp-stem and shoe. C C, C C, represents wrought-iron hoops; D D, key-ways, the keys being used for forcing or driving out the stamp-

stem and shoe from the stamp-head or socket, when
necessary. A represents the stamp-stem, and B the
stem of the shoe.

The stamp, in a broad signification, embraces the tappet already described; the stamp-stem, which is a round
bar of wrought iron, turned and finished; and the stamp-

Fig. 10.

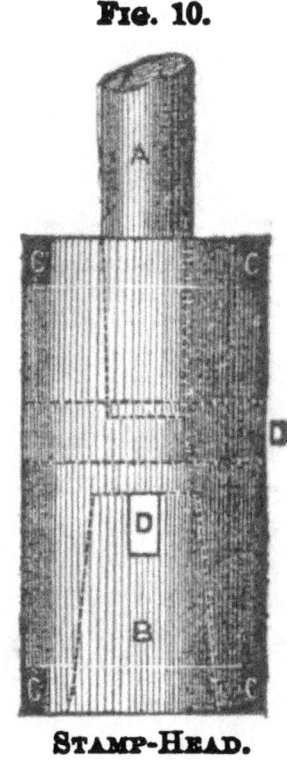

STAMP-HEAD.

head or socket, which is made of the toughest cast-iron, in
the upper end of which the stamp-stem is firmly secured in
a slightly conical opening, and in the lower end of which
is a larger conical opening; into this is fitted with wooden
wedges the stem of the shoe, which may be removed at
pleasure. The shoe should be made of cast-steel; white
iron, however, is usually employed. The round stamp
has almost entirely supplanted the use of the square.

DIES.

The Dies are made of cast steel, or the same quality of iron as the shoes, and are fitted into the bottom of the mortar. These may also be removed at pleasure. Some dies are round, and are fitted into recesses of correspond-

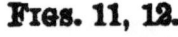

Figs. 11, 12.

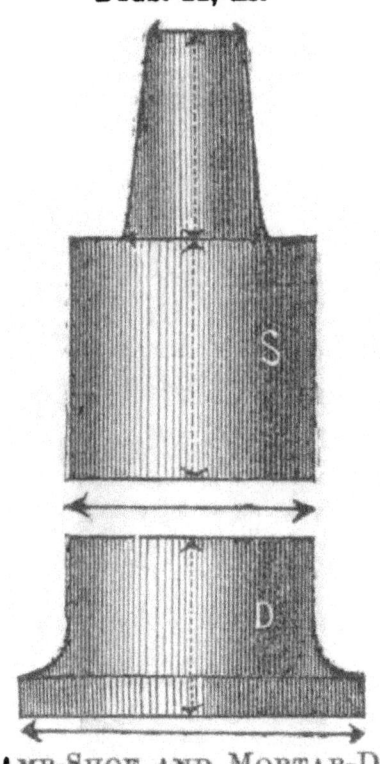

STAMP-SHOE AND MORTAR-DIE.

ing forms in the bottom of the mortar; others are nearly square. There should always be the same number of dies as stamps in a mortar. In Figs. 11 and 12, S denotes the stamp-shoe, and D the mortar die.

SCREENS.

The Screen used for working ores by the *wet* process is made generally of Russia sheet-iron. This iron has a planished, glossy and smooth surface of gray oxide of iron; it should be free from rust or flaws, and be very soft and tough. The severest test of sheet-iron consists in hammering a part of the sheet into a concave form. In the manufacture of this kind of screens, the sheet is perforated by punches varying in size from the number nine to the number one common sewing needle.

The screen used for working ores *dry* is usually made of wire, and varies in fineness from nine hundred to ten thousand meshes to the inch.

GUIDES.

The Guide in which the stamp runs is generally made of the firmest wood, which can easily be obtained, and is secured by bolts to the cross-ties of the battery frame. It is constructed in halves, so that when the stem-way through it, or between its parts becomes worn, it may be refitted to the stem by dressing down its faces.

GRINDERS AND AMALGAMATORS.

Machines in almost endless variety have been employed for the reduction of ores from a granular state to an impalpable powder or slime condition, and for the amalgamation of the precious metals. Of all these the most efficient in use are technically termed " Grinders and Amalgamators." The Grinder and Amalgamator is a large pan provided with grinding plates which operate similar to millstones. The upper plate, called a "muller," revolves, and besides grinding, incorporates the quicksilver into the charge of ore under treatment.

On the Pacific coast, Grinders and Amalgamators are employed almost to the entire exclusion of all other machines invented for like purposes. They take their respective names from the several forms of their grinding surfaces, and may be properly classified as the Tractory, Plane and Conical.

WHEELER AND RANDALL'S EXCELSIOR GRINDER AND AMALGAMATOR.

In the Tractory Conoidal machines, the grinding surfaces, or surfaces of revolution are generated by revolving the tractory curve round its directrix as an axis. The invention embraces two styles of machines—one called the "Tractory Conoidal Grinder and Amalgamator," the other, the "Excelsior Grinder and Amalgamator," represented in the annexed cut. The former of these machines has the greater base of its muller upward, and the latter has it downward. The quicksilver in the former is much less agitated than in the latter, which by many operators is regarded a very valuable desideratum. On the other hand the material to be amalgamated passes direct from the grinding surfaces into the quicksilver, thus precluding, after having been burnished, the possibility of becoming coated with any foreign matter. 'Preference is mostly in favor of the Excelsior.

Fig. 18.

Excelsior Grinder and Amalgamator.

Theoretically, also practically considered, the grinding effect of either the "Tractory Conoidal" or the "Excelsior Grinder and Amalgamator" are fifty per cent. greater than that of the Plane, and thirty-four per cent. greater than that of the Conical Pan of the same size. The reason for this great inequality is that the vertical wear of the tractory-formed grinding-plates is uniform, whereas in the plane and conical pans the more central portions of the grinding-plates impede the wear and consequent grinding effects of the less central portions. The grinding-plates under consideration are subject to the same laws as pivots—in fact are but large pivots—on which subject see "Des Ingenieurs Taschenbuch," pages 143 and 144—"Weisbach Mechanics," pages 313 to 318 inclusive—Professor Rankine's "Manual of the Steam Engine and other Prime Movers," page 17—M. M. Armengaud, sen., Armengaud, jun., and Amouroux's "Practical Draughtsman and Book of Industrial Design," pages 181 to 183—and the supplement of Wheeler and Randall's "Quartz Operator's Hand-Book"—First Edition.

The great popularity of the "Excelsior Grinder and Amalgamator," however, arises not alone from its superior grinding properties, but among other things, from its unequalled stability and convenience of *working* and *cleaning up*. Thus, when it is necessary to put in a new set of shoes and dies, or to clean up, instead of employing fall and tackle, levers and skids, with waste of time and much hard work, as is the case in other pans, the muller of the Excelsior, by means of the screw at the centre, is raised with the utmost facility entirely out of the way of the operator.

The form of the hub and stationary grinding-plates pitching down toward the circumference also greatly facilitate the drawing off the pulp and "cleaning up." When, from stopping the mill or other causes, the pulp in the pan becomes too firmly packed to allow starting the muller by means of the hand-wheel at the side of the pan, it is readily accomplished by removing the key in the centre-screw. The muller, rising, breaks up the pulp. Then, at the pleasure of the operator, it is lowered and secured by the key being returned to its place. For the practical uses of the Excelsior Grinder and Amalgamator, see "Treatment of Silver Ores by Amalgamation —Pan process."

This machine is so constructed that it is readily converted into a continuous Grinder and Amalgamator. Openings on its side at different elevations are prepared for the reception of a discharge-tube, which extends near the centre of the pan, where the material under treatment is lightest and least agitated.

SEPARATORS.

These machines differ in form from each other almost to an unlimited extent. The term "Separator," in general, embraces that class of inventions designed for separating the amalgam and precious metals from their "pulp" or gangue. The machines to which it has come more specially to be applied are large pans, varying from four to say ten feet in diameter, and from one to four feet in depth. A common size is seven feet diameter and two feet deep. If they are made much larger than this, the central portions of the mass being operated on are apt to "pack," while those at the circumference are undergoing proper treatment; or on the other hand, the outer portions become too much agitated, while the inner are being well worked. They are named according to the form of their respective bottoms, to wit : the *Plane*, *Concave* and *Conoidal*.

In the *Plane*, the quicksilver spreading over the entire bottom is apt to be ground with the gangue instead of parted from it.

In the *Concave*, the ore being deepest at the centre where the motion is least, tends to "pack," thus preventing separation from taking place.

To avoid these fatal objections, and otherwise simplify and facilitate the work of the operator, the *Conoidal* was invented, which it is proposed to describe somewhat in detail.

WHEELER AND RANDALL'S CONOIDAL SEPARATOR.

The Conoidal Separator has a curved bottom. The curvature decreases from near the centre towards the circumference, where, with the exception of a bowl and spiral groove for quicksilver, the surface is nearly flat.

By this happy device, uniformity of motion and mixing of the material under treatment are secured; "packing of the pulp" about the centre, or grinding of the quick-

FIG. 14.

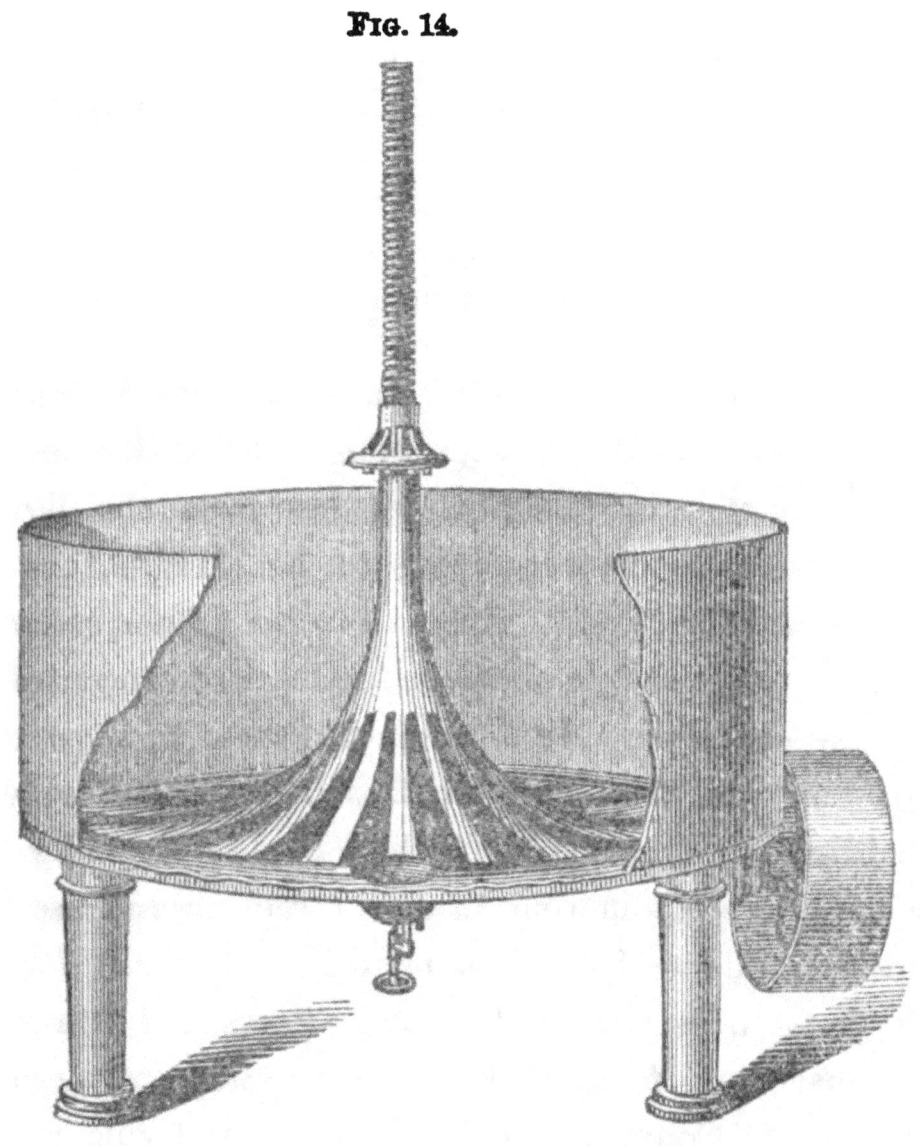

CONOIDAL SEPARATOR.

silver with the gangue, as experienced in other separators, is obviated, and the labor of "cleaning up" is greatly lessened and simplified. This machine is provided with a large screw at the centre, by which the muller can readily be raised entirely out of the way of the operator. It also has the " Patent Self-regulating Quicksilver Discharge Apparatus," which is a very useful and great labor-saving machine. The capacity of the Conoidal Separator is rated at ten tons of ore in twenty-four hours. See " Treatment of Silver Ores by Amalgamation—Pan process."

CONCENTRATORS.

The term " Concentrator" may in general be applied to any device made for separating the heavier portions of granulated or finely-broken ore from the lighter. Generically, concentrators are to be classified as the " dry " and the " wet." Relative to the former, reference is made to the article on the reduction and concentration of ores—dry way. As to the latter, those in most common use are, the Hand or Box Buddles; Revolving Buddles; Jiggers; Dolly Tubs; Sludge Dressing Machines; Tables, differing in size, structure and motion; Cradles; Hides of animals, having the hair on; Riffles; Blankets to be washed by hand; Revolving Blankets; and machines of more recent construction and specially known as " Concentrators." As a general rule, the reduced ore should be carefully sized before being dressed by almost any one of the machines enumerated. On this continent, however, it is not usually done. The sizing may be effected by revolving sieves running in water.

The Concentrators which seem most entitled to consideration are the "Eureka" and the "Tabular."

WHEELER AND RANDALL'S EUREKA CONCENTRATOR.

This machine in some of its features does not essentially differ from the "Sludge dressing machine," quite favorably known in Europe. In comparison, however, it possesses some very important advantages over that machine. Thus it concentrates more rapidly and is continuous working, both as to the dividing or separating of the ores according to their specific gravities, and the discharging of the quicksilver collected. The diameter of this machine is about five feet. The form of the bottom is that of a double-inclined plane, whose upper edge coincides with the diameter, and is at right angles to the horizontal driving-shaft. Its motion is angular and vibratory. It has a central opening, like the "Sludge dressing machine," for the discharge of the gangue. At each of its lower sides is a long narrow opening through the rim, for the discharge of the ore concentrated into the "sulphuret box," whence it is delivered continuously as gathered into the tank below. The invention of this *continuous discharge apparatus* was patented by Mr. Wheeler in 1864, and it is to be regretted that certain *soi disant* inventors are appropriating it to their own uses with unwarrantable freedom. At the circumference of the pan-bottom are depressions for collect-

3*

ing quicksilver, which is conducted off as gathered by means of The Wheeler Patent Self-regulating Quicksilver Discharge Apparatus.

The Concentrator has a conical-formed cover, at the centre of which is a large bowl. Near the circumference of the cover is a circular screen. The material to be treated is run into the bowl; thence flows over the cover through the screen into the pan, and is then disposed of as before described.

THE WHEELER AND RANDALL TABULAR CONCENTRATOR AND AMALGAMATOR.

In the accompanying Fig. 15 of The Wheeler and Randall Concentrator and Amalgamator, A represents the frame; B, head of the table; E, sulphuret-box; I, discharge-pipe; H, bottom of table; L, copper plate, silvered; P, shower-pipe; T, connecting-rod; D, self-discharging quicksilver apparatus; C, receiving-trough, perforated at the bottom and secured to the head B. The bottom of the head-end is about four inches lower than the opposite or discharge-end. The table has a vibratory or pendulum motion at right angles to the current of pulp and water passing through it. The vibrations are three inches in length, and three hundred and sixty a minute. The material, as it comes from the stamps or other reduction works, passes direct into the receiving-trough, thence is showered into the main table. By the quick, short, vibratory motion the material is

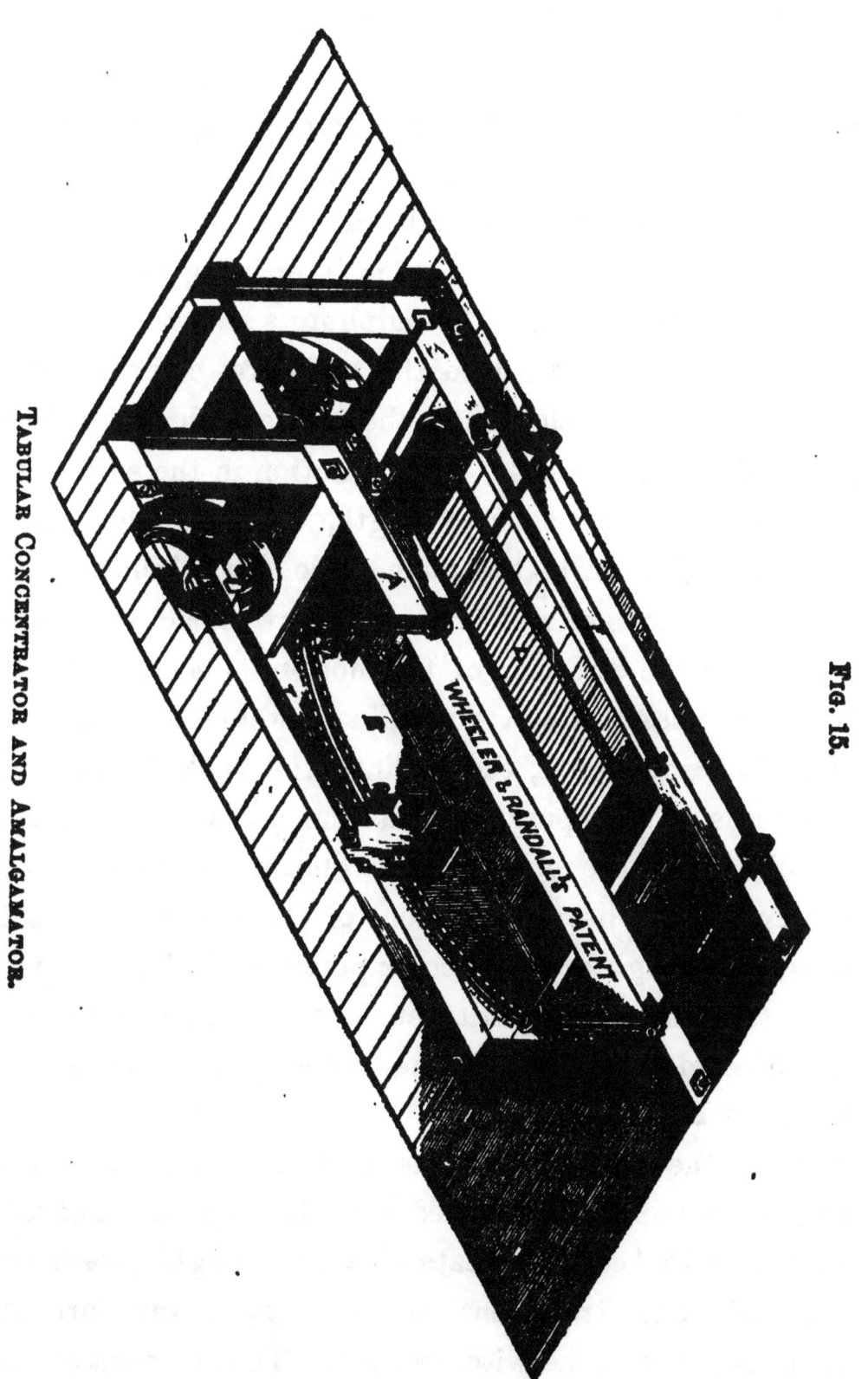

TABULAR CONCENTRATOR AND AMALGAMATOR.

FIG. 15.

kept buoyed up, or prevented from packing; at the same time the precious metals and the heavier portions of the material being treated (almost ever the richer portions of the ores) by their specific gravity, sink to the lowest depths in the table. The substances subject to amalgamation are taken up by the quicksilver lying in the deepest part of the table. The sulphurets and heavier portions of the ores pass into the sulphuret-box through a small opening in the head, about one inch and a half above the table bottom. The agitation in the sulphuret-box, owing to its shorter length, being much greater than the agitation in the main table, the sulphurets and heavier portions of the ores are thrown backward and forward in the direction of the motion to the ends of the box, and pass through a small adjustable opening into the discharge-pipe I, thence into the tank below. So simple is this apparatus that the utmost novice can readily adjust the pipe so as to obtain any desired percentage of the heavier portions of the ores under treatment. The lighter and poorer portions of the ores pass in a thin sheet, being triturated at the same time over the silvered plate L, off at the discharge-end, opposite to the head B of the table.

From the pipe P, water is continuously thrown in a fine shower upon the silvered plate L. The effect of this shower is to keep the plate clean, thoroughly wash the material under treatment, and send back any particles that may have otherwise escaped. To the credit of the machine, however, as reported by the most experienced manipulators of ores, the copper plate and shower pipe scarcely pay the slight expense of their first cost, so effectual has been the operation of the table on the ores prior to their reaching the plate in question. The quicksilver is kept at any desired height in the table by means of the self-discharging quicksilver apparatus D. The surplus quicksilver, as it comes from the reduction works, flows off into a vessel prepared for its reception. The amalgam gathered in the bottom of the table may be "cleaned up" from time to time according to the wish of the operator.

For the treatment of most ores, as a general rule, the weight of water to that of ore should be as four and a half to one. The stream of pulp should flow uniformly into the machine. The capacity of the machine is rated at five tons of ore in twenty-four hours. A much larger quantity per day has been treated in it with entire success. The frame should be well secured by bolts to a good foundation.

RETORTS.

Retorts, with their appurtenances, are vessels employed in " distillation " and " sublimation." The operation is termed " sublimation " if the resulting product is a solid, and "distillation" if it is a liquid or fluid. The operation of vaporizing the mercury of gold and silver amalgam, of conducting away and condensing the vapor, is termed " retorting." Retorts for this purpose are usually made of cast-iron. Their shape and dimensions vary according to the caprice of the constructor as well as the amount of work to be performed.

GOLD RETORTS.

The Gold Retort is a conical-shaped vessel, having its smaller end or bottom usually rounded. The cone is secured to the mouth or larger end by means of a clamp and wedge. The joint between the cone and rim of the retort is made tight by means of loam or clay. The shorter leg of a syphen-formed exhaust-pipe is screwed into and through the cover; the other and longer leg is made to pass through the condenser, which is a simple vessel kept filled with cold water while the retort is in use. Sometimes instead of a vessel as a condenser, cloths kept wet with cold water are wrapped round the pipe. The inside of the retort, before being used, should be rubbed over with chalk, flour, charcoal, or like substance, to prevent the amalgam from sticking. It is not advisable to fill the retort over two-thirds full, lest, by its expansion, the vapor or exhaust-pipe becomes choked, and the apparatus bursts in consequence. The heat at first ought to be gentle, and applied to the shorter leg of the pipe and upper part of the retort; then to all parts of the body of the retort alike. The heat may then be gradually increased, but never raised to a degree much above a bright cherry red. The quicksilver, on being condensed, flows into the receiver, which is generally kept filled with water.

SILVER RETORTS.

The Silver Retort, as it is called, is as well adapted for the retorting of gold as of silver amalgam, when the quantity is large. The most approved retort is the cylindrical. Some of the advantages of this style are, that it can be turned so as to substitute a sound for a burnt side, without destroying the brick-work; also, for a similar reason, it can be replaced by a new one. For example, a retort formed of convenient size and proportions is sixty inches long; the main or cylindric portion on the inside (thickness of metal one and one-half inches,) is twelve inches in diameter and thirty-six inches in length; the neck, of conoidal form, is twenty-four inches long and two and one-half inches least inside diameter. There is a flange on the end of the neck to which the exhaust-pipe is secured by bolts. This pipe, the retort being properly set much like the common horizontal flue boiler, turns downward at right angles, and passes through the condenser. When in use, the condenser is kept filled with cold water to the height of the discharge-pipe near its top, by a small supply-pipe which reaches near its bottom. The quicksilver-receiver, placed under the end of the exhaust-pipe, is kept nearly full of water. The other end of the retort has an angle, flange or hood, into which the door is fitted. Within the cylindric portion of this flange, or hood, are two inclined or helix-formed lugs opposite each other. A " bar," turning upon a pin in the centre of the door, holds it firmly in its place, by the ends of the bar catching under the inclined or screw-formed lugs.   The joint between the door and bottom of the flange is made tight usually by means of clay luting.

Two tiers of amalgam trays may be used in this retort at the same time, one above the other.   The bottoms of the lower tier are circular, so as to conform to the shape of the retort.   When the retort is in use, temporary brick-work is sometimes built up in front of the door to prevent the escape of heat.

CRUCIBLES.

Crucibles are small conical vessels, narrower at the bottom than at the mouth, for reducing ores in docimacy by the dry analysis, for fusing mixtures of earthy and other substances, for melting metals and compounding metallic alloys.   They ought to be refractory in the strongest heats, and not readily acted upon by the substances ignited in them, not porous to liquids, and capable of bearing considerable alternations of temperature without cracking, on which account they should not be made very thick.   The best crucibles are formed from a pure fire-clay, mixed with finely-ground cement of old crucibles, and a portion of black-lead or graphite.   Some pounded coke may be mixed with the plumbago.   The clay should be prepared in a similar way to that for making pottery-ware.   The vessels, after being formed, must be slowly dried, and then properly baked in the kiln.

Black Lead Crucibles are made of two parts of graphite and one of fire-clay mixed with water into a paste, pressed in molds, and well dried, but not baked hard in the kiln. They bear a higher heat than the Hessian crucibles, as well as sudden changes of temperature; have a smooth surface, and are therefore preferred by the melters of gold and silver. This compound forms excellent small or portable furnaces.

INGOT MOLDS.

Ingot Molds are simple troughs made of cast iron. They are slightly wedge-formed, the cavity being somewhat larger at the top than at the bottom. For example, the inside of an ingot mold two inches by six inches at the bottom and two inches in height, should be about two and one-eighth inches by six and a quarter inches at the top. The inside of the mold should be oiled before being used.

It is customary to put the nominal capacity of the ingot mold per cubic inch for gold at one hundred and twenty-five dollars, and for silver at four and one quarter dollars.

STAMP MILL.

The expression "Stamp Mill," in a limited sense, signifies simply the batteries; but in a broader sense, embraces not only the batteries but the prime movers and all the machinery applicable to the reduction and concentration of ores, amalgamation and further treatment of the precious metals.

The work to which stamps are peculiarly well adapted is the reduction of fragmentary ores coming from the rock-breaker to a granular state, so as to pass through No. 4 or No. 5 screens. In this capacity, stamps arranged in straight batteries are unrivalled in efficiency, durability and simplicity. To accomplish the same kind of work, thousands of different machines have been put in operation. These invariably have proved so many failures, and have shortly found their way to the furnace, or been left where first reared as monuments of a sad experience.

A stamp weighing six hundred and fifty pounds, and making ninety twelve-inch drops a minute for twenty-four hours, reduces about two and one-half tons of hard, tough rock from and to the respective sizes named, and requires nearly one-horse power for each ton of rock so reduced. For the reduction of gold and silver ores the usual ratio by weight of water to that of rock is as four and a half to one, while for the Lake Superior copper ores the ratio is as twenty to one by volume.

The feeding of the rock into the battery should be uniform. The practical rule is to so feed that iron to iron shall be heard as often as every tenth or fifteenth fall of the stamp.

Looking to the erection of a Stamp Mill, *location* is a subject of scarcely less importance than that of determining the *prospective* value of the vein or mineral deposit for which the machinery is designed. The site should be selected with a view chiefly to its elevation, its accessibility to the mine and common thoroughfare, and to the facilities offered by the surrounding country for fuel, timber and water. The height of the elevation for immediate mill purposes should not be less than twenty feet; besides, the fall below for carrying off the tailings should be at least one and a half inches to the foot, unless there may be a surplus of water.

The floors of the mill should be arranged in steps so that the material under treatment shall pass from machine to machine through the series with the least handling. To secure the best results the foundations of all the machinery must be firm—those of the mortars in particular. The timber in mortar foundations is placed, as to its grain, either horizontal or vertical. The horizontal foundation is composed sometimes of a single timber, and sometimes of several. In either case they are laid across mud-sills, or on brick or stone work. A single timber, termed a " mortar block," being employed, the mortars are secured in their places either by large bolts or by being let into it. If several sticks (usually three) are used, they are bolted together, one acting as a bed for the mortar and the other two as side clamps. The vertical foundation is constructed in distinct sections, each of which, designed for a single mortar, is composed of several timbers framed and securely bound together with bolts, keys and hoops. In case of their being several sections, they are connected by clamps running lengthwise of the batteries. The bed-rock is usually sought as the base of the vertical foundation, but when this lies too deep, resort is had to mud-sills, brick or stone work. The end of each frame of timbers should be dressed level and smooth so as to make an accurate fit with the planed bottom of the mortar, which is to be bolted to it. The vertical foundation is preferable to the horizontal, for being separate from the battery frame, the jar occasioned by the fall of the stamps is not communicated to the other machinery. It is also firmer, thus rendering the blow of the stamp more effective. Besides, in case of decay it can readily be repaired without detriment to the frame of the battery.

Water and steam being common to most driven machinery, a description simply of the principal modes of their connection with the quartz machinery of Stamp Mills is in this place deemed sufficient. The main driving-shaft is in some instances coupled with the cam-shaft, and in others is connected with it by means of a counter-shaft and gearing, or belts and pulleys. When the coupling is used, the cam-shaft is frequently employed as a driving-shaft for the other machinery. In case of the counter-shaft being adopted, it is usually placed at the foot and in front of the batteries, but sometimes is put from fifteen to twenty feet from them, and ten or more feet above the level of this position. The connection between the intermediate and motor-shafts may be made by direct coupling or by intervening machinery. The stamps, grinders and amalgamators, and commonly the separators are driven by this shaft,

Another counter-shaft for the rock-breaker, and sub-counter-shafts deriving their motions from the main intermediate shaft, for driving respectively the amalgamators and the concentrators, are generally introduced. When the elevated counter-shaft is employed, the rock-breaker is driven by a sub-counter-shaft placed on the top of the battery frame. As a general proposition the application of intermediate shafts is preferable to that of direct couplings; for in the former case, the batteries and the amalgamating machinery may be run or stopped at pleasure, the one without interfering with the other; whereas in the latter case, the amalgamating machinery, receiving its motion through the cam-shaft, is subject to all its delays. Besides, both the motive machinery and that driven from the cam-shaft are exposed to no little wear and tear, arising from the jar produced by the fall of the stamps and the concussion of the cams and tappets. But whether it is better to put the intermediate shaft near the foot of the battery or to elevate it as described, is not fully settled. If it is placed in the lower position its journals and bearings, without extreme care, are liable to be injured by crushed ore from the battery getting into the pillow-blocks. Another important objection is that the cam-shaft belt is rendered too short to secure the best mechanical results; otherwise, as the running machinery is entirely out of the way of the workmen, the plan seems almost faultless. The shaft placed in the upper position, these objections do not apply; but one well worthy of no little attention does: the framework for the running machinery, the belts, tightening pulleys and other appliances taking up much valuable room, restrain the operations of the workmen.

Having described in preceding articles, somewhat in detail, the best known quartz machinery, and discussed in the present article the leading features of Stamp Mills, it is now proposed to consider Silver and Gold Mills, with reference to the arrangement of the machinery and the modes of operating it.

SILVER MILL.

Aside from motors and connecting machinery, the principal machines of a Silver Mill are the Rock-breakers, Batteries, Tanks, Grinders and Amalgamators, Separators, Concentrators, Retorts, Crucibles and Ingot Molds.

Rock-Breakers.—The ore, as it comes in blocks from the mines varying in size from a hen's egg to that as large as a man can well lift, is fed into this machine, which reduces it to an inch or so in diameter. A Blake Stone Breaker (see description) with a feed opening nine by fifteen inches area, the crank making one hundred and eighty revolutions a minute, reduces from one hundred to one hundred and fifty tons of hard rock in twenty-four hours to a size suitable to be fed into the batteries, and requires about six-horse power. The rock, or ore, as reduced, falls upon the battery "feed floor." Prior to the introduction of the rock-breaker into the mines the ore was "spalled" by hand-rollers or by stamps weighing a thousand or more pounds and having a fall of at least three feet. Ore is still prepared for further reduction by hand, when the amount does not exceed fifteen or twenty tons a day.

Batteries.—The Batteries succeed the Rock-breaker at a distance of twelve or more feet. The "feed floor" of the former, depending on the available fall, is ten feet, more

FIG. 16.—CALIFORNIA SILVER MILL.

or less, below the floor of the former. The feed floor should be nearly level with the feed opening in the batteries, so as to save the labor of lifting the ore. It is often from two to three feet below, but no good reasons can be assigned for this plan. The feeding of the ore should be uniform—(see description of Gold Mill). The material fed into the batteries being reduced sufficiently fine to pass through No. 4 or No. 5 Russia iron screens, flows into the tanks. It is customary to estimate that each stamp weighing six hundred and fifty pounds, and making ninety ten-inch drops per minute, will crush about two and one-half tons of hard rock in twenty-four hours, and requires nearly one-horse power for each ton reduced. As a general proposition, all the conditions complied with, this is near the truth; but owing to the lessening weight of the stamp by wear, and the unavoidable hindrances to continuous running, two tons a day for any considerable time may be regarded good work.

Tanks.—The Tanks come next, and are usually set with their tops level with the upper side of the sills of the battery frame. They vary greatly in size. A common size is five by seven feet, and three feet deep. In these the ore is well settled, so that the water runs off quite clear. Proper "settling" of the ores is a matter of vital importance. The tank capacity cannot be too extensive. It is well to arrange these machines in a series, so that the ore borne along by the water through the first of the series may be deposited in the second or third, and so on to the satisfaction of a rigid test. Silver ore, for the most

4

part being very friable to no little extent, becomes so fine in the process of reduction as readily, without extreme care, to pass off in the current of water. While a "charge" is being taken from one tank, the stream of ore and water from the battery is turned into another of the receptacles. The capacity of a tank of the proposed dimensions is between four and five tons of ordinary crushed ore.

Grinders and Amalgamators.—These machines are placed near the tanks with their rims level with the tops of the latter. The "Excelsior Grinders and Amalgamators," with muller four feet diameter, require about five feet fall and seven feet run, including the platform. A less fall, however, will suffice by setting the separator with its top just below the level of the discharge-opening of the machine under consideration. In these machines the ore from the tanks is worked in charges and reduced to an impalpable paste or slime, or "slum"—(see "Treatment of Silver Ores by Amalgamation—Pan process.") An "Excelsior" of the proposed size, the muller making sixty-five revolutions a minute, will reduce full five tons of ore in twenty-four hours, and require about five-horse power.

Separators.—The Separator follows the Grinder and Amalgamator. It is placed sometimes so that one of its sides is under the platform of that machine, and sometimes just from under it and more elevated, as suggested above. The Conoidal Separator, seven feet diameter, requires of itself from four to five feet fall and nine feet run, including platform. As to details for operating this machine, see "Treatment of Silver Ores by Amalgamation—Pan process." In it can be well worked about ten tons of ore a day, at an expense of one horse power, providing the ore is not allowed to pack.

Concentrators.—The Concentrator succeeds the Separator, and requires, in case of the Tabular being employed, about three feet fall and ten feet run. (For a particular description of this machine, see "Tabular Concentrator.") The ore runs from the separator into it. The lighter and poorer portions of the material flow off at the discharge end, and the heavier and richer portions pass off at the sulphuret-pipe into the tank. It is capable of working from five to ten tons of ore a day, and requires not to exceed one half of a horse power.

For description of *Retorts, Crucibles* and *Ingot Molds,* the reader is referred to the articles under their respective heads.

GOLD MILL.

Gold Mills are in far greater variety than Silver, for the reduction of the vein-stone, concentration of the richer portions of the material under treatment, and the amalgamation of the precious metals.

The same machines, that is, the rock-breaker, stamp, grinder and amalgamator, separator and concentrator, are employed both in gold and silver metallurgy, but are subject to greater modifications in arrangement in the treatment of gold-bearing rock than in the treatment of silver ores.

In the treatment of silver, all the ores that are crushed are ground in pans, separated, etc., but in the treatment of gold-bearing rock, modifications in the arrangement of the machinery are made according to the character of the gold,—viz.:

1st. If the gold is very fine and uniformly diffused through the rock, the machinery as arranged under the head of "Silver Mill" can be employed to advantage for its treatment; except that it is advisable to employ a greater number of concentrators and amalgamators. The first series of these should be used immediately after the batteries, but simply as amalgamators. The whole mass should be treated.

2d. If the gold in the rock is contained mostly in the

sulphurets, the concentrators and amalgamators should be employed immediately after the batteries. The concentrated portions of the rock should be ground in the grinders and amalgamators, worked in separators, and finally treated in another series of concentrators.

3d. If the gold is quite coarse and clean in the rock, it will not be advisable to employ grinders and amalgamaters, and separators.

4th. If the gold is coarse and coated it is advisable to work the rock by the " continuous process," as it is termed.

Modifications in machinery, as observed, must necessarily be made according to the character of the gold and the rock.

The weight of authority is, in wet crushing, in favor of amalgamating in the battery. Great care in feeding the quicksilver into the battery must be observed.

If too little quicksilver be fed the amalgam will be dry and granular, and will readily pass off in the stream of crushed ore. If too much quicksilver be fed, the amalgam will be liquid, and will readily flow off in the current.

The quicksilver should be fed in small quantities into the batteries, and fed often. It had better be fed by being sprinkled through buckskin, or some other porous substance.

The rule to be observed is, to so feed the quicksilver that the amalgam passing the screens may be indented by a gentle pressure of the finger, and yet so firm as to

retain the indentation. Copper-plate, silvered, is sometimes employed both inside and outside of the battery,
but its use seems not well warranted. Its highest claim
as an amalgamator rests almost entirely on accident;
but though it cannot be relied on as safe, it at least is
harmless and cheap, and since the true principle is to
"catch" gold at all points where it can be done profit-
ably, copper-plate, as an auxiliary, may perhaps be
employed sometimes with seeming advantage.

REDUCTION AND CONCENTRATION OF ORES—DRY WAY.

The ore, broken to the size of hickory nuts, maize, or so, is reduced by batteries (see Dry Mortar), rollers, or grinders. The grinding-plates are usually of iron, but sometimes of stone.

To fit the ore for concentration it should be granulated—rather than pulverized. At the same time, in most cases, it has to be rendered quite fine to become detached from its vein-stone.

Silver and Gold ores, for the most part, require finer reduction than do those of Lead, Zinc and Iron. To accomplish fine granulation, none of the machines spoken of are free from objections. Thus the batteries tend to pulverize, so do the grinders; while the faces of the rollers soon wear uneven and suffer portions of the ore to pass too coarse between them. On the whole, however, for this class of work, rollers properly constructed, arranged and cared for, are perhaps the least objectionable.

In further fitting the ore for concentration, it is to be thoroughly *dried* and *sized*. The *drying* is effected by the material being spread on heated plates, passed through or over heated cylinders, let fall through currents of hot air, or otherwise exposed to the action of heat.   The *sizing* is accomplished by means of sieves or wire-cloth bolts of different degrees of fineness.   These put in motion, the prepared ore is delivered into the sieve or bolt of finest mesh; thence its unscreened portions pass into the next coarser bolt, thence into the next coarser, and so on, through the series.   The tailings are returned to the reduction works.

Machines of various devices for concentration, or separating the heavier portions of ore from the lighter, have been employed—some involving the well-known principle of *winnowing*—some that of projecting horizontally from an elevated position, the material to be separated, into the air, which acting as a resisting medium causes the ore to fall from the point thrown at distances corresponding to the different specific gravities of its granules having the same size and form—the heavier being the more remote,— and others, that of sending puffs of air up through shallow layers of the granulated ore, causing its grains to become arranged vertically according to their respective specific gravities; when by means of mechanical appliances their separation is completed.   So far as appears, the successful application of puffs of air to ore-dressing was first made by Mr. Thomas J. Chubb, of New York.   The writer has operated the Chubb Concentrator on the ores of lead— galena, with highly satisfactory results, and is credibly informed that it has been worked on the ores of graphite, nickle, cobalt, silver, gold, and iron with equally flattering success.   The intensity of the puffs of air is to be regulated according to the fineness and specific gravity of the material under treatment. Thus the puff barely of sufficient strength for the coarser portions of the ore is found much too intense for the finer, which goes to prove the absolute necessity of careful sizing.

The subsequent treatment of ores prepared in the *dry* way is the same as that of those reduced in the *wet*, that is, they are ground, amalgamated and separated in the same manner—(see Pan process). As to the relative merits of the two modes of reduction and concentration, it is to be stated that larger yields of the precious metals may generally be obtained from the same amount of ore prepared by the *dry* way than by the *wet*. The expense of labor and the destruction to life and machinery are however greater, so that the *wet*, as before observed, is almost universally adopted—at least on this continent.

EXAMINATION OF MINERALS.

The behavior of a mineral or metallic substance when subjected to certain tests, such as being heated in glass tubes, either closed or open, on charcoal, in the oxidizing flame, melted in glass fluxes on platinum wire, treated by acids and other reagents, determines its kind. When borax is employed, it should be vitrified. The vitrified borax is prepared by melting at a brown-red heat, in a clay or Hessian crucible, common borax into a clear transparent glass. It is preferable that the platinum wire should have a small eye or loup at its test-end. The spirit-lamp should be employed. To make the borax-test, heat the test-end of the platinum wire to redness, dip it into the powdered borax, then heat again, bring the borax globule in contact with the hot specimen to be tested, and apply the flame till the specimen is dissolved. The color or other characteristics will show what kind of mineral is under examination. The specimen, if metallic, with scarce an exception, should be thoroughly roasted and its oxide subjected to the test.

Borax (biborate of soda). Before the blow-pipe (B B,) intumesces and fuses to borax-glass; with fluor-spar and bi-sulphate of potash, colors the flame green; soluble in water, the solution changes vegetable blues to green.

Fluor-Spar (fluoride of calcium) B B, decrepitates and fuses to an enamel; the flame continued, the specimen assumes a cauliflower appearance; heated with salt of phosphorus in a glass tube, it etches or roughens the inside of the glass.

Nitre—B B, deflagrates vividly, detonates with combustible substances, dissolves readily in water; not altered by exposure.

Phosphorus moistened with sulphuric acid, and heated, gives a green tinge to the flame.

Chlorine—A substance containing chlorine, combined with the salt of phosphorus and oxide of copper, on the platinum wire, colors the flame deep blue.

Quartz (silicic acid) undergoes no change alone; with soda, readily fuses with effervescence to a transparent glass.

Sulphur heated in an open glass-tube emits fumes of sulphurous acid; heated with soda, the compound, moistened with water, blackens silver.

Alumina (sesquioxide of aluminum) B B, unaltered, both alone and with soda, fuses with borax with great difficulty, also with salt of phosphorus; moistened with cobalt solution and brought to a high heat, becomes blue; is not attacked by acids.

Antimony (oxide) B B, readily melts, evaporates in white inodorous fumes; colors the flame greenish-blue; with borax, in oxidizing flame, forms a yellow glass, which becomes colorless on cooling; in reducing flame the globule contracts, becoming gray and vitrious; with carbonate of soda, reduces to metal; is very friable and brittle.

Arsenic (oxide) B B, volatilizes in white fumes of garlic odor; heated to redness, burns with a pale bluish flame.

Bismuth (oxide) B B, evaporates, leaving a yellow coating, on charcoal; color, on wire, hot, is dark brown— cold, is yellow; with borax, is colorless; with carbonate of soda, on charcoal, reduces to metal.

Lime (oxide of calcium) B B, infusible alone; with borax, effervesces; with comparatively large quantity of borax, forms a clear glass which becomes angular on cooling; in the flame of the oxyhydrogen blow-pipe emits a most dazzling white light, and fuses at the edges.

Copper (oxide) B B, in the oxidizing flame, fuses; in the reducing flame, forms metal; with borax, in the oxidizing flame, colors the glass green; in the reducing flame, brown-red.

Cobalt (oxide) B B, unchangeable by itself; with carbonate of soda, on charcoal, forms a gray magnetic powder; with borax, both in oxidizing and reducing flame, gives a deep-blue bead.

Tin (oxide) B B, in the oxidizing flame, presents a dirty-yellow color; with carbonate of soda, in reducing flame, on charcoal, reduces to metal; with borax, forms a clear glass.

Lead (oxide) B B, in oxidizing flame turns first blue, then fuses to a glass of orange color; with carbonate of soda, on charcoal, in the reducing flame, reduces to metal; with borax, forms glass yellow while hot, colorless when cold.

Nickle (oxide) B B, infusible alone; in the oxidizing flame, with borax, forms an orange-red globule which becomes nearly colorless on cooling; in the reducing flame, on charcoal, the bead becomes gray; in the reducing flame, with soda, on charcoal, reduces to a magnetic powder.

Zinc (oxide) B B, in the oxidizing flame, exhibits a whitish-green color; while hot, this oxide is slightly yellow—when cold, is white; with borax, forms glass which in an intermittent flame becomes milky; in the reducing flame, on charcoal, reduces to metal, which readily sublimes.

Iron (pre-oxide) B B, in the oxidizing flame, is unchanged; in the reducing flame, blackens and becomes magnetic; with borax, in the oxidizing flame, forms glass bright-red while hot, pale dirty-red when cold; in the reducing flame, forms glass varying from bottle-green to black-green; with carbonate of soda, on charcoal, reduces to metal as a dark magnetic powder.

Manganese (oxide) infusible alone, becomes brown by heat; with borax, in the oxidizing flame, much oxide employed, the glass is black; little oxide employed, the glass is of an amethyst color—in the reducing flame, and on charcoal, this latter globule becomes colorless and so remains if quickly cooled; with soda, in the oxidizing flame and on platinum foil, forms an opaque green glass.

Silver (oxide) B B, with borax, in oxidizing flame, forms a white opaque glass; in reducing flame, with carbonate of soda, readily reduces to metal.

Gold (metal) B B, with carbonate of soda, in the reducing flame, readily fuses; the most appropriate test is by cupellation.

Tellurium (oxide) B B, colors the flame green, fuses and sublimes; with borax, in the oxidizing flame, forms a colorless glass; in the reducing flame, the glass becomes gray; its behavior with carbonate of soda is similar as with borax; on charcoal, readily reduces to metal.

Soda (oxide of sodium) B B, colors the flame deep-yellow.

BEHAVIOR OF SOLUTIONS OF METALLIC OXIDES WITH REAGENTS.

ZINC (OXIDE).

Potassa and Ammonia, in neutral solutions, give white gelatinous precipitate, readily soluble in an excess of the precipitants.

Oxalic Acid occasions precipitates which are soluble in acids and in fixed alkalies.

Ferrocyanide of Potassium produces a white gelatinous precipitate, insoluble in hydrochloric acid.

COBALT (OXIDE).

Potassa produces a blue precipitate, insoluble in an excess of the precipitant, but soluble in carbonate of ammonia; the precipitate becomes green by exposure to the air, and dingy-red when boiled.

Alkaline Carbonates produce a red precipitate, which upon being boiled becomes blue.

Ferrocyanide of Potassium gives a green precipitate which gradually turns gray.

Cyanide of Potassium precipitates cyanide of cobalt as a brownish-white precipitate from acid solutions, soluble in excess of the precipitant, and not again precipitated by acids.

IRON (PROTOXIDE).

Potassa and Ammonia produce a floculent precipitate of hydrated protoxide of iron which at first is nearly white, but which readily becomes colored by exposure to the air.

Alkaline Carbonates produce a white carbonate, gradually becoming colored, though not so readily as the oxide. It is soluble in sal-ammoniac, but a colored precipitate makes its appearance by exposure to the air.

Ferrocyanide of Potassium produces a white precipitate if the air be excluded; but by exposure to air, or by contact with nitric acid or chlorine, it forms Prussian-blue.

Phosphate of Soda produces a white precipitate, which after a time becomes green.

IRON (SESQUIOXIDE).

Potassa and Ammonia produce a voluminous reddish-brown hydrate, insoluble in excess. The precipitation is prevented by the presence of organic acids, sugar, etc.

Alkaline Carbonates throw down precipitates of rather a lighter color.

Sulphocyanide of Potassium changes the color of neutral or acid solutions to a deep blood-red color, owing to the formation of a soluble sulphocyanide of iron. This is the most unerring test, as it will detect the slightest trace of iron, even in the most dilute solution.

Phosphate of Soda produces a white precipitate, which

becomes brown, and finally dissolves on the addition of ammonia.

MANGANESE (OXIDE).

Ferrocyanide of Potassium produces a pale-red precipitate, soluble in free acids.

Ferricyanide of Potassium gives a brown precipitate, insoluble in free acids.

Phosphate of Soda produces a white precipitate, persistent in the air.

LEAD (OXIDE).

Potassa and Ammonia produce white precipitates, soluble in a great excess of potassa, but insoluble in ammonia.

Alkaline Carbonates produce white precipitates, soluble in potassa.

Phosphate of Soda occasions a white precipitate, soluble in potassa.

Oxalic Acid produces in neutral solutions a white precipitate.

Hydrochloric Acid and Soluble Chlorides produce a heavy white precipitate, soluble in boiling water, out of which it separates, on cooling, in brilliant crystals. This precipitate is soluble in potassa.

Sulphuric Acid and Soluble Sulphates produce a white precipitate, sparingly soluble in dilute acids, but soluble in solution of potassa, and assuming a black color when moistened with hydrosulphuret of ammonia.

SILVER (OXIDE).

Potassa and Ammonia produce a light-brown precipitate, readily soluble in ammonia; ammoniacal salts prevent precipitation by potassa.

Carbonated Alkalies produce a white precipitate, soluble in carbonate of ammonia.

Phosphate of Soda produces, in neutral solutions, a. yellow precipitate, soluble in ammonia. Solution of ignited phosphate of soda gives a white precipitate.

Protosulphate of Iron produces a white precipitate, consisting of metallic silver.

Hydrochloric Acid and the *Soluble Chlorides* produce a white curdy precipitate, even in exceedingly dilute solutions. This precipitate becomes violet, and finally black, without, however, suffering decomposition by exposure to light. It is insoluble in diluted acids, but readily soluble in ammonia. When heated it fuses, without decomposition, into a horny mass.

A Bar of Metallic Zinc precipitates silver from its solution, in the metallic state.

MERCURY (OXIDE).

Potassa produces a yellow precipitate, insoluble in excess. If an insufficient quantity of the alkali be added, the precipitate is reddish brown. The presence of ammoniacal salts gives a white precipitate.

Carbonate of Ammonia produces a white precipitate.

Phosphate of Soda produces a white precipitate.

Ferrocyanide of Potassium produces a white precipitate, which eventually becomes blue, owing to the formation of Prussian blue.

Ferricyanide of Potassium produces, in solutions of nitrate and sulphate, a yellow precipitate; in solutions of the chloride, *none*.

Iodide of Potassium produces a cinnabar red precipitate, soluble in excess. It crystallizes out of a hot solution in magnificent crimson spangles.

BISMUTH (OXIDE).

Potassa and Ammonia produce a white precipitate, insoluble in excess.

Alkaline Carbonates and Phosphate of Soda produce white precipitates.

Ferrocyanide of Potassium occasions a white precipitate, insoluble in hydrochloric acid.

Ferricyanide of Potassium produces a light-yellow precipitate, soluble in hydrochloric acid.

Iodide of Potassium produces a brown precipitate, readily soluble in excess.

COPPER (OXIDE).

Potassa produces a voluminous blue precipitate, which, by boiling, loses water and becomes black.

Ammonia, added in small quantities, produces a green basic salt, which dissolves in excess, forming a fine blue solution. In this solution, potassa produces in the cold, after a time, a blue precipitate, which, by boiling, becomes black.

Carbonate of Potassa produces a greenish-blue precipitate of basic carbonate of copper, which, by boiling, is converted into black oxide.

Phosphate of Soda produces a greenish-white precipitate, soluble in ammonia, forming a blue solution.

Ferrocyanide of Potassium produces, even in very dilute solutions, a chocolate brown precipitate, insoluble in dilute acids, but decomposed by potassa. This is a very delicate test.

Ferricyanide of Potassium produces a yellowish-green precipitate, insoluble in dilute acids.

Cynanide of Potassium produces a yellowish-green cyanide, soluble in excess of cyanide of potassium.

Metallic Iron, when introduced into solutions of oxide of copper, becomes covered with a deposit of reduced copper.

ANTIMONY (OXIDE).

Potassa and Ammonia produce a white precipitate, partially soluble in excess of the reagent.

Alkaline Carbonates and Phosphate of Soda behave in a similar manner.

Metallic Zinc throws down metallic antimony as a black powder; if nitric acid be present the sesquioxide is precipitated at the same time.

TIN (OXIDE).

Potassa, Ammonia and their carbonates produce a

white precipitate, soluble in potassa but not in ammonia; by repose, and more rapidly when boiled, the solution is decomposed—*metallic tin*, peroxide of tin and potassa being formed.

Phosphate of Soda produces a white precipitate.

Ferrocyanide of Potassium gives a white gelatinous precipitate.

Ferricyanide of Potassium occasions a white precipitate, soluble in hydrochloric acid.

A Bar of Metallic Zinc precipitates tin in small grayish-white metallic spangles.

Chloride of Gold produces a purple precipitate, a mixture probably of peroxide of tin and metallic gold (purple of Cassius), insoluble in hydrochloric acid.

PLATINUM (PEROXIDE).

Potassa produces a yellow crystalline precipitate, consisting of the double chloride of platinum and potassium; the addition of hydrochloric acid favors its formation; it is insoluble in acids, but dissolves with the aid of heat in potassa; it is very slightly soluble in water, and insoluble in strong alcohol.

Subnitrate of Mercury produces a yellowish-red precipitate.

Chloride of Tin, in presence of free hydrochloric acid, communicates to solutions of bi-chloride of platinum a deep red-brown color, without producing any precipitate.

Platinum soluble in aqua-regia.

GOLD (PEROXIDE).

Potassa (in heated solutions), after a time produces an inconsiderable reddish-brown precipitate, consisting of teroxide of gold mixed with terchloride of gold and potassa.

Ammonia produces a yellow precipitate (aurate of ammonia or fulminating gold).

Ferrocyanide of Potassium communicates to solution of gold an emerald-green color.

Protosulphate of Iron produces in concentrated solutions an immediate dark-brown precipitate of metallic gold. In dilute solutions a blue coloring is first perceived, followed by a brown-colored precipitate.

Oxalic Acid produces a precipitate of metallic gold.

Protochloride of Tin, to which a drop of nitric acid has been added, communicates a reddish purple color to very dilute solutions; in concentrated solutions a red-purple precipitate (purple of Cassius) is formed.

A Bar of Metallic Zinc precipitates metallic gold in the form of a brown coating.

Gold is soluble in aqua-regia.

TELLUROUS ACID.

The caustic alkalies and their carbonates produce a white precipitate, soluble readily in potassa, and soluble in alkaline carbonates.

Sulphurous Acids, Alkaline Sulphites, and Metallic Zinc, produce a black precipitate of metallic tellurium.

Protochloride of Tin and Protosulphate of Iron produce a black powder, which, on being rubbed, assumes a metallic lustre.

Tellurium is soluble in nitric acid.

BLACK FLUX.

Black Flux is prepared by introducing gradually in small quantities into a crucible heated to a very dull redness, a mixture of either two parts of cream of tartar and one of nitre, or equal parts of cream of tartar and nitre.

WHITE FLUX.

White Flux is similarly prepared to Black Flux, except that the mixture consists of one part of cream of tartar and two parts of nitre.

BLOW-PIPE.

The Blow-pipe is an instrument used for directing, by a current of air, the flame of a lamp or candle upon a mineral substance to fuse or oxidize it. The flame consists essentially of two parts—the "oxidizing" and "reducing." 1st. The "oxidizing" part is the outer and slightly luminous flame; 2d. The "reducing" part, which is hottest, is the inner blue flame.

The "reagents" mostly used in making blow-pipe tests are charcoal, carbonate of soda, cyanide of potassium, and borax. Black Flux and White Flux are much employed in assaying.

Caution.—Cyanide of potassium is a very dangerous flux. Its compound with a "nitrate" or "chlorate" explodes with great violence.

Charcoal performs the part of a cupel as well as that of a reagent. The best charcoal is made of young soft wood.

The "cupel," or "support," consists of a sound piece of coal, sawed or broken lengthwise, having a small cavity made in its plane side near the edge, to hold the substance to be tested. The borax is prepared by being melted or vitrified, and pulverized.

A.—Blow-pipe Assays of Silver Ores containing Sulphur and Arsenic:

1st. Roast the pulverized ore within a shallow cavity

on the coal support. To do this, direct by the blow-
pipe the oxidizing flame, that is the extreme point of the
outer flame, upon the powdered ore; turn the specimen
occasionally so as to expose all of its parts to oxidation.
Let the blast at first be gentle, then increased and continued till the sulphurous odors cease.

2d. Pulverize the roasted ore, and put it upon the
charcoal within a prepared cavity; cover it over with
carbonate of soda, and direct upon it the reducing flame,
that is, the inner blue flame; let the blast at first be
gentle, and chiefly employed in fusing and bringing
together all parts of the flux, then increase the blast,
shaking the support or coal, so as to gather the metal in
one globule, which cool and extract from the slag.

3d. Cupel the button thus obtained, adding to it three
parts of lead in a boneash cupel. To do this, employ
the oxidizing flame by using a very gentle blast. Cool,
weigh the button, and compare its weight with that of
the original ore.

*B.—Place within the cavity, on the charcoal, a mixture
of:*

Powdered ore, (not necessarily roasted)..... 1 part
Borax-glass (pulverized)................... 1 part
Lead: (Litharge preferable when earthy sub-
    stances are present; and then it is also
    advisable to make an addition of carbon-
    ate of potash and glass).............. 3 parts
and direct upon it the reducing flame till all the metal is

5

gathered into one button.   Let the blast at first be gentle and chiefly directed upon the borax, then increased and continued till the object is attained.   Extract and cupel the button as in example *A*.

C.—Blow-pipe Assay of Horn Silver.   Place within the cavity, on the charcoal, a mixture of:

Horn Silver (chloride of Silver)........... 1 part
Carbonate of soda (in excess)............. 3 parts

and reduce in the inner blue flame to a button, etc.

ASSAY.

Assays are three kinds—*Mechanical, Dry,* and *Humid.*

MECHANICAL ASSAYS.

Mechanical Assays consist in washing, or otherwise freeing, without the aid of chemical agents, the metallic substances from sand and other impurities. "Panning out," or separating the gangue (earthy matter) from the metallic substances by washing in common mining pans, also winnowing, as practiced on rich dry sands are familiar examples of mechanical assays and require no explanations—the former of which often furnishes safer and more practical data for extensive operations, especially in gold mining, than either the *dry* or *humid* way.

DRY WAY OF ASSAY.

The "dry" way of assaying usually requires the addition of fluxes to the ores to facilitate their reduction, and separate the metals so reduced from the gangues.

Assay of Galena, or Ores of Lead containing Sulphur.—Introduce into an earthen crucible in the following order:

Powdered Ore.......................... 10 parts
Iron in strips or plates............... 1 to 3 parts
Black Flux............................ 30 parts

Place upon the black flux a thick layer of salt, and on this a piece of charcoal. Put an earthen cover on the crucible, and fuse its contents, letting the temperature, at first low, be gradually increased to bright red, and continued at this degree about half an hour. Gently rap the crucible, to settle all the metallic particles; then cool the assay, and extract the button of lead.

Assay of Oxidized Ores of Lead.—Introduce into an earthen crucible:

Powdered Ore........................... 10 parts
Carbonate of Soda................. 30 to 40 parts
Pulverized Charcoal...................... 3 parts
Iron, in strips or plates, in case any sulphur
        may by present...................... 1 part

Cover with salt, etc., and conduct the assay as in the preceding case.

Assay of Iron.—Fuse in a covered crucible about one hour, a well triturated mixture of:

Powdered and Roasted Ore.............. 2 parts
Fluor-Spar................................ 1 part
Charcoal.................................. 1 part
Common Salt, *e. g.* (for covering).......... 4 parts

Extract and weigh the button of cast iron thus obtained. Various other fluxes, as lime, clay, etc., may be employed. The nature of the ore and its gangues determine the kind and proportion of fluxes to be employed.

Assay of Copper Ores, containing no other metals besides Iron and Copper.—Heat gradually at first, in an earthen crucible, and afterwards increase the heat to bright red, which continue fifteen minutes:

Powdered Ore.............................. 1 part
Black Flux................................. 3 parts

Extract from the slag, and weigh the button of copper thus obtained.

Assay of Copper Ores, containing Sulphur, but otherwise similar to the above.—Fuse in an earthen crucible, at a dull red heat, equal parts of the powdered ore and dried borax, extract from the slag the matte (crude copper) button, which pulverize; roast slowly in an earthen crucible, and stir in the meantime with a steel rod till sulphurous acid ceases to be evolved; then increase the temperature to a white heat, which continue for several minutes; next, mix in the same crucible:—

Roasted matte........................... 1 part
Black Flux, from..................... 3 to 4 parts

Cover the mixture with a layer of fused borax, and subject it to a cherry heat for twenty minutes in a wind furnace; then extract and weigh the button of copper.

Assay of Copper Ores containing Arsenic and various other metals.—Obtain and pulverize the matte as in the preceding case, then roast with it powdered charcoal till the garlic odors of arsenic cease to be exhaled. Reduce the matte thus obtained as in the last case with black flux and borax. Cupel the button in a bone-ash cupel, with pure lead. Throw a little borax-glass over the globule when its rotation ceases and brightening occurs; cool, and weigh the button of copper.

Assay of Gold or Silver or Gold or Silver Ores.—Fuse in an earthen crucible:

Powdered Ore........................... 4 parts
Litharge............................... 4 parts
Black Flux............................. 3 parts

If the ores contain much oxide of lead, add only black flux. If the ores are very rich in pyrites, add litharge and nitre. If the button obtained be an alloy, for instance, of gold, silver, copper and lead, make additions to it of silver and lead, so that the prepared alloy shall contain as near as may be of—

Gold .... 1 part
Silver............................... 3 parts
Lead, from........................... 12 to 16 parts

First fuse the lead in a bone-ash cupel, within a muffle; then add the gold and silver inclosed in a piece of paper, and continue the heat till the button brightens and becomes tranquil; cool, and weigh the button. To separate the gold from the silver, called " parting of gold," anneal, beat the button into a thin plate, make it into a roll, which is termed a "cornet." First heat this plate or cornet in dilute nitric acid as long as the acid acts upon it, then in concentrated nitric acid till all of the silver is dissolved. Thoroughly wash, dry and ignite the cornet. The weight of silver is equal to the weight of the button before parting, less that of the refined cornet.

REMARK.—Chloride of silver (horn silver) cannot be decomposed by heat alone. It fuses at a temperature of 500° F. At a temperature of ebullition it is decomposed by caustic potassa and soda, and if a little cane sugar be added, the silver will be reduced to the metallic state. It is also reduced to the metallic state when fused with the alkalina and earthy carbonates. Thus one part of the chloride of silver fused with two parts of carbonate of soda is reduced; also one part of the chloride of silver with two parts of chalk and two parts of charcoal reduces to the metallic state. Chloride of silver cannot be decomposed by charcoal unless water be present; in which case hydrochloric acid and carbonic oxide are formed, and the silver liberated. Copper, tin and lead reduce chloride of silver in the dry way. Mercury partially decomposes it, forming an amalgam of silver. The chloride of silver, in a state of fusion, is very destructive to earthen crucibles, scarcely less so than litharge.

HUMID WAY OF ASSAY.

Assay of Galena.—1st. Digest the powdered ore in equal parts of nitric acid and pure water.—2d. Filter and digest the residuum several hours with a strong solution of carbonate of soda.—3d. Filter and digest the second residuum in dilute nitric acid, and again filter.—4th. Add either a solution of the sulphate of soda or sulphuric acid to the collected filtrates, as long as any precipitate takes place.—5th. Filter, then wash and dry the residuum.—6th. Reduce the residuum with powdered charcoal in an earthen crucible; cool, and weigh the button.

Assay of Copper Ores.—1st. Digest the powdered ore in dilute nitro-muriatic acid.—2d. Filter the solution.—3d. Add ammonia in excess to the filtrate.—4th. Filter and wash residuum in ammonia.—5th. Evaporate the filtrate to dryness.—6th. Dissolve the dried filtrate in muriatic acid.—7th. Add clean iron or zinc plates to the solution much diluted.—8th. Wash dry, and weigh the copper precipitate.

The Swedish Method.—1st. Digest the thoroughly roasted and powdered ore in concentrated muriatic acid, stir well, and evaporate nearly to dryness.—2d. Add, slowly, strong sulphuric acid till the odor of hydrochloric acid is no longer perceived; then heat till the vapors of sulphuric acid begin to develop.—3d. Dilute with distilled water, and keep the assay warm for some time.—4th. Filter and wash the residuum in warm water.—5th. Add and retain in the solution clean zinc plate till no reddish tinge be given to a clean plate of iron dipped into the assay.—6th. Wash thoroughly the precipitate, dry at a temperature of 212°, and weigh the metal.—7th. In case the precipitate be found partially oxidized, complete the oxidation by heating the assay in an open dish, and weigh the oxide of copper; thus—100 : 79.8 : : weight of oxide : weight of copper.

Assay of Silver Ores.—1st. Digest the pulverized ore in nitric acid.—2d. Add muriatic acid or solution of common salt to the silver solution as long as any precipitate takes place.—3d. Filter and dry the residuum.—4th. Reduce the dry residuum with carbonate of soda, in an earthen crucible; then cool, and weigh the button of silver.—*Chloride of Silver*, called, as found native, "horn-silver," is completely insoluble in nitric acid, also in water.—*Ammonia*, boiling solutions of the chlorides of potassium, sodium, barium, strontium and calcium, also cyanide of potassium, in solution, dissolve the chloride of silver. Concentrated sulphuric acid slowly decomposes it. Iron and zinc at the ordinary temperature decompose it, especially in presence of free muriatic acid.—*Bromide of Silver* in almost every respect resembles chloride of silver; it is, however, less soluble in ammonia.—*Iodide of Silver*, found also native, is readily converted into chloride of silver by muriatic acid.

Assay of Gold Ores.—1st. Digest the pulverized ores in one part of nitric and four parts of hydrochloric acid. —2d. Dilute, filter and evaporate the filtrate almost to dryness, to expel excess of acid.—3d. Dilute the dried filtrate in pure water; then boil the solution with a solution of sulphate of iron, which precipitates the gold as a dark purple powder.—4th. Filter and heat the residuum with hydrochloric acid.—5th. Filter, wash, dry and weigh the gold powder. Oxalic acid, substituted for the sulphate of iron, precipitates the gold in large flakes.

ASSAY OR ANALYSIS OF IRON ORES CONTAINING MANGANESE.

1st. Digest the roasted and pulverized ore in dilute hydrochloric acid.—2d. Filter, wash residuum, and add washings to the filtrate.—3d. Add muriate of barytes until no further precipitation takes place.—4th. Filter, wash, and add washings to the filtrate.—5th. Evaporate the filtrate nearly to dryness, and to it add sufficient nitric acid to transform the sulphate of iron to peroxide.—6th. Add solution of caustic ammonia in excess to the solution.—7th. Filter and reduce the iron to the magnetic state by heating the residuum with resin in an iron crucible; then cool and weigh.—8th. Precipitate the oxide of manganese from the filtrate, by expelling the excess of ammonia by heat.—When the ores contain much alumina or silex, flux them with three or four times their weight of caustic potash, then digest in hydrochloric acid, and proceed as above.

ASSAY OR ANALYSIS OF ORES CONTAINING GOLD, SILVER, COPPER, LEAD, IRON AND SULPHUR.

1st. Digest well the pulverized ore in nitric acid.—2d. Filter, wash residuum (1), and add washings to filtrate (1).—3d. Add to filtrate (1) hydrochloric acid or solution of common salt, which precipitates the silver as a chloride.—4th. Filter, and digest residuum (2) in hydrochloric acid.—5th. Filter, wash residuum (3) in warm water, and to filtrate (3) with the washings add filtrate (2).—6th. Reduce the chloride of silver with carbonate of soda by fusion, and weigh the button of silver.—7th. Add to filtrate (3), sulphate of soda in solution, which precipitates the lead as a sulphate.—8th. Filter and add the residuum to residuum (1).—9th. Evaporate filtrate (4) to any desirable extent.—10th. Add, in excess, to concentrated filtrate (4) ammonia, which precipitates sesquioxide of iron.—11th. Filter, wash residuum, and add washings to filtrate (5).—12th. Dry and heat the residuum in hydrogen gas within a glass tube as long as any vapor of water is disengaged, then weigh the iron. This powder, with glass as a flux at a high heat, becomes a button of iron.—13th. Treat filtrate (5), after evaporating it to dryness with hydrochloric acid, then add clean iron or zinc plates to the solution diluted. Wash, dry, and weigh the copper precipitate.—14th. Treat residuum (1) first with a strong solution of carbonate of soda, then with dilute nitric acid; and to the combined filtrates add sulphuric acid, or a solution of the sulphate of soda. Wash, dry, and reduce the precipitate with powdered charcoal in an earthen crucible; then cool and weigh the button of lead.—15th. Digest the last residuum in nitro-muriatic acid; add chloride of sodium in solution, filtrate, precipitate the gold from its solution by the addition of sulphate of iron in solution; wash, dry, and weigh the gold.—16th. If the gold may be alloyed with silver and copper, precipitate the copper from the last filtrate by the addition of iron or zinc plates; wash, dry, and add the weight of the precipitate to the copper already obtained. Heat the residuary ore in a strong solution of chloride of sodium, filter and precipitate the silver with a clean copper plate; wash, ignite and add the silver to that already obtained.—17th. Burn off the sulphur and weigh residuary ore. The sum of the weights of the gold, silver, copper, lead, iron and calcined ore taken from the weight of the original ore leaves the weight of the sulphur.

ROASTING.

Roasting is employed to dissipate the volatile parts of the ore by heat, and is effected in *Heaps* or in *Furnaces*.

HEAPS.

Alternate layers of fuel and ore, usually as it comes from the mine, are heaped up to the depth of several feet. The lowest or ground layer is of wood, arranged by cross piling so as to afford a free circulation of air. The upper layers may be of wood or coal. The ratio of fuel in volume to that of ore varies from 1 to 6, to 1 to 18.

Fine ores, and those rich in sulphur, require less fuel than coarse ores and those poor in sulphur. The fire is kindled through vertical openings or chimneys, which extend to the ground layer. These openings are closed when the fuel has well taken fire. The roasting should be slow and uniform in all parts of the heap. The heat may be regulated by opening or closing the draft-holes and chimneys. Several days, and even months, sometimes are required for roasting one heap. Ores similarly piled with fuel are sometimes roasted in walled inclosures provided with side openings.

FURNACES.

Furnaces are in great variety. Those mostly approved for the roasting of ores, embracing also calcining and chloridizing, are the reverberatory. The interior walls of the furnace should be of the best fire brick, laid edgewise; the outer walls may be of common building brick or stone. The furnace must be well tied with iron rods, and carefully dried before being used.

Reverberatory Furnace.—The Reverberatory Furnace is constructed sometimes with one and sometimes with two hearths or soles, one above the other. In the "double-hearth" furnace, in the treatment of silver ores, the "roasting" and "sulphatization" are effected on the upper sole, and the "calcining" and "chlorodizing" on the lower. The ore pulverized fine is charged upon the upper sole to the depth of two to four inches, and is kept well stirred during the "roasting." The heat should be at a low temperature, not exceeding brown or dull red. The access of air should be free. A small jet of steam let into the furnace assists in regulating the temperature, and also facilitates oxidiation. The addition of powdered charcoal in small quantities may be made to advantage when the ores contain arsenic. If the ores are poor in sulphur, add from two to three per cent. of the sulphate of iron. The first operation of "roasting" and "sulphatizing" is accomplished in four or five hours. Then, through an opening in the upper hearth, the ore is let fall upon the lower, where it is heated for some time at a temperature not much higher than that mentioned above. The heat is then gradually increased to cherry-red, at which it is kept during the time required for "calcining" and "chloridizing." The heat should never exceed bright red. The ore is frequently stirred. When calcination is complete, a mixture of common salt, melted and pulverized, and seven parts of cold calcined ore are added to the hot ore, estimated at fifteen parts, and quickly and thoroughly mixed with it by stirring. Calcination is usually effected in four or five hours, and chlorination in fifteen or twenty minutes.

SMELTING.

Smelting is a process by which ores are reduced to the metallic state. It is performed in furnaces of which there are several kinds. The "Backwoods hearth," or "low" furnace, without artificial draught, also the "Scotch and American hearths," or "low-blast" furnaces, are employed on some ores, especially those rich in galena. The furnaces mostly approved for smelting purposes are the "Cupola" and the "Reverberatory," already mentioned under the head of "roasting." Each of these is subject to various modifications in size and form to meet the requirements of the nature of the ore, and of local causes.

The Cupola, or "high-blast" furnace, has a vertical opening of greater height than breadth. This opening or chimney-like shaft in a furnace of medium size is about fifteen feet high, and has an area through its greatest horizontal section of nearly twelve square feet: this section is usually rectangular. However, in the Castilian furnace it is circular, and in the MacKenzie patent furnace, very favorably known among practical furnacemen, it is quite eliptical. The blast, or a strong current of air, is forced into the lower part of the shaft through one or more tongues. The feed-opening is several feet above the bottom. The ore fed into the "Cupola" furnace with the fuel is in contact with it, and subject to the direct action of combustion. The flames, as they rise, impart much of their heat to the ore, arranged in layers one above the other, and finally pass off at a temperature far below that of the flames from the "Reverberatory." The "Cupola," compared with the "Reverberatory," as to first cost, expense for fuel, manipulating and repairs is the more economical; but as to results, especially in the treatment of rich ores containing little or no quartz, the "Reverberatory" has the preference.

In smelting operations, as heat alone is frequently insufficient to induce the necessary chemical changes, fluxes are employed. These are substances having an affinity for some portion of the material under treatment, and either effect its volatilization or form with it a fusible compound or slag. This slag, lighter than the metal, is often separated from it, either by being drawn off from the surface, or suffered to remain until the metal is drawn from underneath. The flux to be employed depends upon local causes and upon the nature of the ores and gangues to be operated on.

Quartz, constituting the gangue of the ore, carbonate of soda, lime, and the oxides of the base metals may be employed as fluxes. The gangue, consisting of the different earths and the oxides of base metals, carbonate of soda, litharge, and quartz may be used. The ores containing much sulphide may have as fluxes, carbonate of soda, lime, metallic iron and litharge. Granulated lead, lead ore and litharge may be regarded fluxes for gold and silver. The fluxes of most general use are carbonate of soda, limestone or lime, litharge, iron and fluor-spar.

GOLD.

Gold occurs in nature chiefly in the form of alloys, the most important of which is "native" gold (so called), consisting of gold, silver, and small quantities of copper, iron, and other metals. Gold is found alloyed with rhodium and paladium; amalgamated with mercury, and occasionally combined with tellurium in the form of an ore, the telluride of gold.

TREATMENT OF ORE CONTAINING GOLD.

The treatment of ore containing gold is effected by three distinct systems—*Smelting, Amalgamation, and Chlorination.*

BY SMELTING.

The method by smelting in most cases is identical with that in the treatment of silver ores, to which reference is here made.

Fuller's Lead Process.—An air-tight vessel is partly filled with lead kept in a melted state by a fire below. The pulverized ore is made to pass down a central pipe or cylinder into the lead, through which it passes, and then rises into the chamber, floats on the surface of the metal, and is removed through a side opening. The gold is supposed to be thoroughly amalgamated or alloyed with the lead, from which it is afterwards obtained by cupellation.

BY AMALGAMATION.

Arastra.—1st. The arastra is a machine employed for the reduction of ores and the amalgamation of the precious metals. It is composed of a stone-paved bottom from six to twenty feet in diameter, wooden-rim two feet and upwards in height, and an upright wooden shaft having two or more horizontal arms, to which are attached, by chains, the "mullers," consisting of large blocks of granite.—2d. The ore in pieces not larger than hens' eggs is introduced into the machine, and the muller set revolving at the rate of four to seven feet a second. As the reduction proceeds, quicksilver from time to time is sprinkled through a cloth or other porous substance into the mass under treatment.—3d. When the reduction and amalgamation is completed, the "slime" or "gangue" is washed off, the amalgam is "cleaned up," "squeezed" and "retorted."

Pan Process.—1st. The rock as it comes from the mines is crushed wet by stamps to a fine granular state, and run into large tanks.—2d. Charges of the reduced ore, with sufficient water to form a thin paste, are thoroughly ground in iron pans. As gold in rock exists almost without exception in a metallic state, friction with a moderate degree of heat is usually found quite sufficient to fit it for amalgamation. Heat, however, is often employed in the amalgamation of gold, and for the most part in the form of steam. The steam is let in some cases direct into the charge in the pan, but is usually applied in steam chambers underneath the pulp. The latter method is preferable to the former. Exhaust steam, charged with oily matter, is detrimental to amalgamation, and hence should not be introduced into the pulp. It is the exhaust steam that is commonly employed for heating the ore under treatment. That a high degree of heat facilitates the amalgamation of gold is not without its advocates. The results, however, of an extended series of experiments made by several expert amalgamators in California were not in favor of its application.—3d. Quicksilver is ordinarily added to the pulp as the pans commence running. To avoid grinding the quicksilver excessively, the addition is sometimes made with the muller slightly raised, after the reduction of the ores.—4th. The charge is then drawn off and washed, leaving the amalgam in the separators.—5th. The proportions usually observed, for instance in the Wheeler & Randall Grinders and Amalgamators, are, ore to the charge (4 feet muller), 1,500 pounds; quicksilver to the charge of ore, 100 pounds; revolutions of muller, 60 to 75; time of reducing, usually about 3 hours.

As gold-bearing rock is seldom found sufficiently rich to render it advisable to treat the entire mass in pans, the above method is subject to various modifications, of which the following are a few:—1st. The heavier and richer portions of the rock, as crushed, are concentrated by revolving-blankets, buddles, concentrators, or other machinery, then pulverized and amalgamated in pans.—2d. Amalgamation is commenced in the batteries during the crushing operation, and is carried on through a series of shaking tables, riffles, and copper plates. The richer portions of the tailings are then concentrated and treated in pans.—3d. Grinding and amalgamating are effected in pans while the reduced ores are flowing continuously through them.—4th. The sulphides or concentrated tailings are roasted in a reverberatory furnace before being ground and amalgamated.—5th. Thin layers of the concentrated sulphides or tailings are spread in inclosures open to the sky, and allowed to remain a long time, for instance a year. The tailings are occasionally turned with shovels, and the lumps broken so as to expose as much surface as possible to the action of the air. Common salt, mixed with the tailings, assist in their oxidation. When thoroughly oxidized they are treated in pans.

The process of amalgamation is not well adapted to raw ores containing the compounds of sulphur, arsenic, tellurium, bismuth, iron, lead, copper, antimony and zinc. These substances not only cause a "sickening" of the quicksilver, but so tarnish or "coat" the gold as to ren der amalgamation next to impossible. For the most part such ores should be thoroughly roasted and the deleterious substances volatilized before attempting amalgamation; or in many cases it may perhaps be better to subject them to smelting.

Hagan's Process.—In this process, pyritous ores are heated red hot in a tall furnace and subjected to the action of a current of super-heated steam, obtained by passing dry steam through a coil of pipe in the fire-place on box below. It is claimed that the heated watery vapor is decomposed, and that the oxygen and hydrogen combine with the elements of the ore, converting the sulphides into oxides with the liberation of the gold, while the sulphur combines with the hydrogen, forming sulphide of hydrogen, which is burned. The wasted ore is reduced to a fine powder or slime, and subjected to amalgamation. The reports are conflicting as to results.

BY CHLORINATION.

1st. Pulverized ores containing gold, having been well roasted, cooled and moistened with water, are put into closely-covered wooden cisterns whose bottoms are so constructed that chlorine gas can permeate the mass from underneath. The ore containing "lime" and "talc" should be roasted with salt, as those substances are highly detrimental to chlorination.—2d. Chlorine gas, produced by heating sulphuric acid, peroxide of manganese and common salt in a leaden generator, is caused to enter the cisterns at the bottom, through leaden pipes. The chlorine gas should be forced through clean water, to cleanse it of muriatic acid, which in several ways is very injurious to the operation. The effect of the chlorine on the gold is to produce terchloride of gold.—3d.

Pure water, after the chlorine has done its duty (which takes from ten to fifteen hours), the covers of the dissolving cisterns being removed, is added sufficient to fill the cisterns even with the ores. The effect of the water is to dissolve the terchloride of gold; the solution is then drawn off into glass vessels.—4th. Sulphate of Iron in solution is used to precipitate the gold, which may then be gathered, melted, and run into bars.

SILVER.

Silver occurs in nature under many forms, viz:

Native Silver.—Native Silver sometimes occurs almost chemically pure, but often is in alloy with gold, copper and other metals.

Antimonial Silver and Bismuth.—These alloys, owing to their scarcity, are of but little commercial value.

Native Amalgam.—The composition of several known varieties of native amalgam is—silver, 34.8; mercury, 65.2; silver, 26.25; mercury, 73.75; silver, 86.49; mercury, 13.51.

ORES OF SILVER.

The ores yielding the greater part of the entire annual product of silver, and hence the most important, are:

Silver Glance.—This ore, known also as Vitreous Silver, Sulphide of Silver, is, on account of its abundance and richness, the first in importance. Its composition is—silver, 87.04; sulphur, 12.96.

Stephanite.—This ore, known as Brittle Silver Ore, Black Silver, ranks second in importance. It is composed of—silver, 70.4; antimony, 14.0; sulphur, 15.6.

Pyrargyrite.—This ore, its name implying ruby silver, usually black; sometimes, however, approaching coch- red. Its composition is—silver, 58.98; antimony, sulphur, 17.56.

Chloride of Silver.—This ore, known as Horn Silver, Muriate of Silver, when found in large quantities, as in Nevada, is very valuable as an ore. It has, however, not been found to extend downward in mineral veins to any considerable depth, but to have been succeeded by sulphides, arsenides and compounds of antimony. Its hardness is but little above that of talc, often the same. Its composition is—silver, 75.33; chloride, 24.67. Many other ores of silver occur, but are comparatively of little commercial importance. The following comprise nearly or quite all the ores of silver known:

Silver Native—Usually found alloyed with copper, gold, etc.

Silver Glance—Vitreous Silver, Sulphuret of Silver (AgS), S12.96, Ag87.04, H=2—2.5, G=7.196—7.365.

Hessite—Telluric Silver, AgTe=Te37.23, Ag62.77, H=2.—2.5, G=8.3—8.9

Naumannite—AgSe=Se27, Ag73, H=2.5, G=8.

Eucairite—Seleniuret of Silver and Copper, CuSe+AgSe=(Cu,Ag) Se31.58, Cu25.26, Ag43.16.

Stromeyerite—Argentiferous Sulphuret of Copper, CuS+AgS=(Cu,Ag) S=S15.92, Ag52.71, Cu30.95, H=2.5—3, G=6.2.

Antimonial Silver—Antimoniate of Silver, Ag^{4}Sb=Sb23, Ag=77, also Ag^{6}Sb=Sb16, Ag83.4, H=3.5—4, G=9.44—9.80.

Arsenic Silver—Consists of Arsenic, Iron, Antimony and Silver.

Flexible Silver Ore—Ferrosulphuret of Silver.

Sternbergite, $AgS + 2Fe^2S^3 = S34.21$, $Ag32.83$, $Fe32.96$, $H = 1.—1.5$, $G = 4.215$.

Miargyrite, $AgS + SbS^3 = S21.35$, $Sb42.79$, $Ag35.86$.

Pyargyrite—Dark red Silver Ore, Ruby Silver, Black Silver, $3AgS + SbS^3 = S17.56$, $Sb23.46$, $Ag58.98$, $H = 2.—2.5$, $G = 5.7—5.9$.

Proustite—Light red Silver Ore, $3AgS + AsS^3 = S19.46$, $As15.16$, $Ag65.38$, $H = 2.—2.5$, $G = 5.422—5.560$.

Frieslebenite—Antimonial Sulphuret of Silver $(PbS + SbS^3) + 2$, $(3Pb,AgS) + SbS^3 = S18.77$, $Sb27.72$, $Pb30$, $Ag22.18$, $H = 2.—2.5$, $G = 6.—6.4$.

Polybasite, $9 (Ag,Cu) S + (Sb,As) S^3 = S17.04$, $Sb5.09$, $As3.74$, $Ag64.29$, $Cu9.93$, $H = 2.—3$, $G = 6.214$.

Stephanite—Brittle Silver Ore, Brittle Silver Glance, Black Silver, $6AgS + SbS^3 = S15.6$, $Sb14$, $Ag70.4$, $H = 2.—2.5$, $G = 6.269$.

Xanthokon (Yellow Powder) $H = 2$, $G = 4.2—4.3$, $3 (AgS + AsS^5) + 2 (3AgS + AsS^3) = S21.36$, $As13.49$, $Ag64.18$.

Bismuth Silver—Bismuth Silver Ore. Composition variable:

Eg. { Bi27, Pb33, Ag15, Fe4.3, Cu0.9, S16.3.
     60.1 10.1   7.8   As2.8

Horn Silver—Muriate of Silver, Chloride of Silver, $AgCl = Cl24.67$, $Ag75.33$, $H = 1.—1.5$, $G = 5.552$.

Iodic Silver—Iodite. Composition variable.

Bromic Silver—Bromite, $AgBr = Br42$, $Ag58$, $H = 1.—2$ $G = 5.8—6$.

Embolite, $2AgBr + 3AgCl = Ag (Cl,Br) = Ag66.96, Br$ 19.84, Cl13.20, H$=2$, G$=5.79 - 5.81$.

Selbite—Carbonate of Silver, $AgC = AgO84, CO_216$. Very soft. Sometimes called Blue Silver.

TREATMENT OF SILVER ORES.

The treatment of Silver ores is effected by three distinct systems, viz.: Smelting, Solution and Amalgamation. Each of these systems is severally subdivided, and subject to almost indefinite modifications.

BY SMELTING.

A radical feature in nearly every method of smelting silver ores is the employment of lead in one or more of its forms. If lead in some form does not occur in the ore to be operated on, it has to be added. In connection with the smelting of silver ores, among the more important properties to be observed of lead and its compounds are as follows, to wit:—1st. The sulphide, oxide or sulphate of silver fused with lead, the silver is reduced to the metallic state, and forms an alloy with the excess of lead.—2d. The sulphate of silver fused with the oxide or sulphate of lead, argentiferous lead results.—3d. Lead has much less affinity for the base metals than for silver. —4th. A fused alloy of lead and silver acted on by a blast of air, the lead becomes oxidized, while the silver remains unchanged.—5th. In the furnace, the sulphide or sub-sulphide of lead and the sulphate of lead react upon each other and produce metallic lead, which takes up the silver, if any be present.—6th. One part of sulphide of lead, and three parts of sulphate of lead, react upon each other and produce litharge.—7th. Sulphide of lead and metallic iron, copper, antimony or zinc in a state of fusion react upon each other, and produce on one hand metallic lead, and on the other the sulphide of iron, etc. If the sulphide of lead at the same time contain the sulphide of silver, the same reactions take place, and the lead and silver enter into alloy, while the iron or other of the metals named is converted into a sulphide. Iron, it may be observed, is usually employed as a flux or reagent. It seems proper here to remark that the treatment of an ore containing silver and gold, inseparable in smelting, is not modified on account of the occurrence of the latter metal.

IN REVERBERATORY FURNACE.

Corinthian Process.—1st. The ore, for example, argentiferous galena of pure quality, is roasted by a slowly increasing temperature until the sulphate thus produced is to the sulphide in the proportion of their equivalents. —2d. The temperature being increased and the charge thoroughly stirred, the sulphide and sulphate of lead reacting upon each other produce metallic lead and sulphurous acid gas, of which the former flows from the furnace into vessels prepared for its reception, the latter escapes by the chimney.—3d. In connection with the reduction of the metal is formed a subsulphide of lead.

This and the sulphate of lead in the proportion of their equivalents react upon each other and produce a further amount of lead.—4th. The roasting is continued until litharge is finally formed by the reactions of one part of sulphide and three parts of sulphate of lead. Charcoal is then added, which reduces the litharge and sulphate of lead in excess to the metallic state.—5th. The alloy, if sufficiently rich in silver, is subjected to cupellation; but if poor in silver, may well first be enriched by the Pattinson process.

IN CUPOLA FURNACE—CASE I.

Silesian Process.—1st. The raw ore, argentiferous galena, and iron-flux prepared by being broken into small pieces and uniformly mixed, are so thrown with the fuel into the furnace that they shall occupy the back of the shaft while the fuel shall occupy the front.—2d. Sufficient blast being applied to induce a gradual fusion, and the reactions of the sulphur and iron together with oxygen taking place, the furnace hearth becomes filled with the rich, liberated lead, and floating slags.—3d. As occasion requires, the lead ready for cupelling is drawn off through an aperture in the bottom of the furnace, and the slags through a more elevated opening. The richer portions of the slags, or properly matte, composed of the sulphide of iron and lead, with a little silver, are subjected to the further treatment of roasting and fusion.— 4th. The proportions of a charge are as follows: galena in small fragments, 100 parts; cast iron, 12 parts;

slag from iron forge, 14 parts; fuel (mineral coal), 126 parts.

IN CUPOLA FURNACE—CASE II.

Rammelsberg Process.—1st. The ores, argentiferous galena, abounding in earthy and metallic impurities, are roasted either in Reverberatory Furnaces or in Heaps in the open air.—*In the Reverberatory Furnace*: The roasting is conducted in the ordinary manner to the desired extent of oxidation. The heat then being increased, the furnace closed and the charge fused, the silicates of lead, of lime, etc., are formed, which with the oxide, sulphate, and sulphide of lead are drawn upon the floor of the furnace, cooled and broken to sizes best suited to treatment in a cupola furnace, *e.g.*, size of oranges.—*In Heaps:* 1st. Ores abounding in the sulphides of iron, zinc and copper being intimately mixed with galena—arranged in order of the sizes of the blocks, the largest at the bottom, are piled several feet high upon a thick layer of wood, and covered with pulverized roasted ore which prevents the access of too much air. The sulphides constitute the greater portion of the fuel. From four to six months are required for the first roasting of a heap containing one hundred and fifty tons of ore. The metallic products of the operation are subjected to a second and third roasting, or rather to a succession of roastings, until the sulphur and other volatile substances are well dissipated.—2d. The charge consisting of roasted ore, silicious slags, litharge and fragments of old cupels—all reduced to the size of hens' eggs and thoroughly mixed, are introduced with the fuel into the furnace. A strong blast applied, inducing fusion and the necessary chemical changes, the lead, matte and slags sink upon the hearth and are disposed of as in Case ı. The proportions of a charge are: roasted ore, 140 parts; highly silicious slags, 40 parts; litharge, 1 to 2 parts.

IN CUPOLA FURNACE—CASE III.

1st.—Argentiferous carbonate of lead, white lead ore, is prepared for the furnace by being broken into small pieces—or occurring as powder—by being made with clay into orange-sized balls and dried. The charge consisting of the prepared ore, litharge, cupel bottoms, and other fluxes suited to the gangues, is thrown with the fuel into the furnace—the charge more on the side of the back wall, and the fuel on the side of the front.—2d. As the blast is let on, and the ore-mixture fused, the carbon of the fuel reduces the free oxides of lead; also those in combination. The metallic lead and slags are drawn off at proper intervals.—3d. The remunerative portion of the carefully-sorted slags are returned for fusion. The fumes collected in the condensing apparatus, and sometimes amounting to nearly half the lead originally treated, are made into orange-sized balls, dried and fused again. The lead obtained is either first treated by the Pattinson process, or cupelled at once for its silver.

IN CUPOLA FURNACES.

Mexican Process.—1st. The ore, composed chiefly of the sulphide of silver, sulphide of iron and quartz, free from the compounds of lead, being intermixed with charcoal of half its weight, is roasted in walled inclosures open to the sky. An equal amount of dry wood is sometimes used instead of charcoal, and the roasting effected in circular kilns measuring on the inside four and a half feet in diameter, and the same in height, built of sun-dried brick, and capable of holding, besides the fuel, about a ton of ore each.—2d. The roasted ore and fluxes being well mixed are introduced with charcoal into the furnace in the following proportions: roasted ore, 75 parts; native carbonate of soda, 16 parts; lead slags, 16 parts; cupel bottoms, 16 parts; matte of previous operations, 25 parts; litharge, 100 parts; charcoal, 50 to 75 parts. As fusion takes place, the metal is drawn off from time to time and cast into ingots, ready for cupellation. The matte that is gathered is returned to the furnace as one of the fluxes for succeeding charges of ore.

Pattinson Process.—1st. This process is founded on these facts: If a melted alloy of silver and lead is stirred while cooling slowly, crystals of lead form and sink, which may be removed with a drainer. A large portion of the lead may thus be separated from the silver.—2d. Cast-iron pans, capable of holding about five tons each, provided with fire-places, are arranged in a series, as A, B, C, D, E, F, G, in a straight line.—3d. The metal of ores containing silver and lead as it comes from ordinary smelting works, is melted, for instance, in pan D, and then allowed to cool very slowly. The metal, while cooling, is stirred, especially near the edges of the pan, with an iron bar. As soon as crystals form and sink to the bottom they are taken out with an iron drainer raised to a temperature somewhat higher than that of the metal bath. From one-half to two-thirds of the charge is thus removed to pan E, and the balance taken to pan C. Other charges of D are similarly treated and disposed of. The charges of C and E are treated and disposed of in like manner, except that the crystals of E go to F, and the balance to D, and the crystals of C go to D, and the balance to B. Thus after successive meltings and drainings, the alloys rich in silver pass to A, while the lead, almost entirely deprived of silver, goes to G. The alloys obtained in pan A are then subjected to cupellation.— 4th. The lead of an alloy treated by this process often contains less than one dollar in silver to the ton. The silver of the enriched alloy should not exceed six hundred dollars to the ton.

Parke Process.—1st. Lead containing silver is fused in large cast-iron pots. Melted zinc is added and well stirred in the alloy. The fire being withdrawn from under the pot, the whole is left at rest for a short time. —2d. The silver and zinc, separating from the lead, form an alloy which is skimmed from the surface of the metal-bath as long as it rises.—3d. The scum alloy containing some lead is heated in a liquation retort. The silver and lead fuse, and to a great extent, flow into prepared moulds. The alloy thus run off is then cupelled; the alloy of zinc and silver remaining in the retort are partially separated by distillation. The silver thus obtained is freed of its impurities by cupellation.—4th. The proportions are: charge of argentiferous lead to the pot, from 6 to 7 tons; charge of zinc to the ounce of silver, by estimation, 1.5 to 2 pounds; quantity of silver to the ton of lead, 10 to 15 ounces; time of stirring alloy after the addition of zinc, 10 to 15 hours; silver to ton of alloy prepared for cupellation, 1,000 ounces.

Liquation Process.—1st. This process is founded on these facts: Lead and copper fused together form an alloy which, if rapidly cooled, maintains an intimate admixture, but, if slowly cooled, separates. An alloy of lead and copper, slowly heated to near its point of fusion, also separates; silver, if contained in the alloy, goes with the lead.—2d. Either an alloy of copper or silver, or matte (crude black copper, reduced, but not refined from sulphur and other impurities) containing silver, as it comes from the smelting furnace, is melted with lead of about four times its weight, in a cupola furnace, and cast into plain circular plates which are suddenly cooled. These plates, called "Liquation Cakes," are arranged on their edges with alternate layers of charcoal, in a liquation furnace. The charcoal is then ignited, and a degree of heat produced somewhat below that of the fusing point of copper. The lead and silver melt and flow into a receiver, while the copper, in a porous state, retains the forms of the original cakes. If the separation may have been imperfect the cakes are further treated by being raised to a higher degree of heat in the sweating furnace. The silver is then separated from the lead by cupellation.

BY SOLUTION.

Augustin Process.—1st. This process is founded on the solubility of chloride of silver in a hot concentrated solution of common salt.—2d. The ores are crushed dry by stamps, and pulverized in suitable mills.—3d. The roasting in a reverberatory furnace is commenced at a low temperature with a free access of air. By careful, uniform roasting, at a dull red heat, the sulphates of the various metals are produced. The heat is then increased to cherry red, which decomposes the sulphates of iron and other base metals, but not the sulphate of silver.—4th. Common salt, previously melted, pulverized and mixed with cold ore, is added to the hot ore in the furnace and thoroughly mixed with it by stirring. The sulphate of silver is thus converted to chloride of silver.—5th. The apparatus for the humid operations consists of a large heating reservoir, a series of dissolving tubs, two large settling cisterns, four precipitating tubs to each one of the dissolving tubs, and two large receptacles, arranged in the order here given, on descending steps. The dissolving and precipitating tubs are nearly cylindrical. They are provided with filters made of small sticks and straw, covered with cloth. A vertical partition resting on the filter divides each tub into two unequal compartments.—6th. The chloridized ore being put into the larger compartments of the dissolving tubs, sufficient of the hot salt solution from the heating reservoir above, to completely immerse the ore, is let into the tubs; they are then left at rest one hour. The discharge-cocks of the heating reservoir and tubs then being opened, the hot salt solution is filtered through the contents of the tubs, and run off from the smaller compartments, at openings at first above the level of the ore, afterwards at openings near the bottom of the tubs, into the settling cisterns, until a test with clean copper shows no trace of silver in the filtered solution.—7th. Copper (copper cement) is put into each of the upper two precipitating tubs in the several series of four, and iron (wrought scrap-iron) into each of the lower two. The chloride solution from the settling cisterns is then slowly filtered through the several series of precipitating tubs, and the filtered solution run into the large receptacles below. The silver is precipitated by the copper in the upper tubs, and the copper in solution, if contained by the ore at first, is precipitated by the iron in the lower tubs. The silver is taken twice a week from the precipitating tubs and refined. The filtered solution in the receptacles is pumped into the heating reservoir and used again.—8th. The proportions usually observed, are: ore before roasting should contain of sulphur not less than 20 per cent.; charge of ore to the furnace, for roasting and calcining, 500 pounds; charge of melted salt, 35 pounds; roasted ore, cold, mixed with salt, 220 pounds; time, roasting on upper sole of furnace, 4 to 4½ hours; calcining on lower sole of furnace, 4 to 4½ hours; time of chloridizing, from 15 to 20 minutes; degrees of heat of salt solution, 131° Fahr.; time of dissolving and precipitating, 20 to 24 hours; solution of salt, run through each tub to one thousand pounds of ore, 200 to 250 cubit ft.; depth of copper in precipitating tubs, about 6 inches; depth of iron in precipitating tubs (if the ore contains copper worth saving), 6 inches.

REMARK.—Zinc, antimony and arsenic are detrimental to the Augustin and Ziervogel processes.

Ziervogel Process.—1st. This process is founded on the solubility of sulphate of silver in hot water.—2d. The ore, as in the Augustin process, having been thoroughly pulverized, is carefully roasted and calcined till the sulphates of iron and other base metals are completely decomposed, but none of the sulphate of silver, which is much more obstinate. When small quantities of the roasted ore thrown hot into water give only a very slight blue color, the calcination is regarded complete.—3d. The sulphatized ore is treated in all respects the same as the chloridized ore in the Augustin process, except that pure water is employed instead of solution of salt. The degree of heat of the water for dissolving should be 149° Fahr.

Von Patera's Process.—1st. In this process the ores are thoroughly pulverized and chloridized by roasting with common salt.—2d. Hot water, to dissolve the chlo-

rides of various base metals, is filtered through the chloridized ores put in tubs similar to the dissolving tubs in
the Augustin process. The ores are then cooled and
transferred to similar but smaller tubs.—3d. Hyposulphite of soda, in cold solution, is then filtered through the
ores and run into precipitating tubs until all the chloride
of silver is completely dissolved.—4th. Polysulphide of
sodium, sufficient to produce a neutral liquor, is then
added, which precipitates the silver as a sulphide in sacks
fitted to the inside of the tubs. (Polysulphide of sodium
is produced by fusing common soda ash with sulphur, and
subsequently boiling the product, dissolved in water, with
sulphur in a finely divided state.) This neutral liquor,
the hyposulphite of soda, is preserved for lixiviating purposes.—5th. The sulphide of silver thus obtained, after
being washed in warm water, pressed and dried, is heated
under muffles, with free access of air, till nearly all the
sulphur is expelled. The metallic silver is then refined.

Freiberg Process, by Sulphuric Acid.—1st. The matte,
impure copper containing silver, as it comes from the
furnace is finely pulverized, thoroughly roasted and chloridized as in the Augustin process.—2d. The chloridized
matte, put into tubs, is subjected to the action of hot sulphuric acid. The copper, with what iron may be present,
is dissolved, while the silver is but slightly affected.—3d.
The solution of the sulphate of copper is drawn off into
vats and crystallized.—4th. The undissolved remainder,
containing the silver, is smelted with lead and cupelled.

Rammelsburg Process, by Sulphuric Acid.—1st. Granulated copper containing silver placed in tubs is subjected, in connection with abundance of air, to the action of hot sulphuric acid trickled over it.—2d. The copper, thus converted into the sulphate of copper, runs off in solution through a series of troughs, in which it is deposited in the form of rough crystals. This liquor is returned, heated, and again trickled through the material under treatment.—3d. The rough crystallized salt is washed, dissolved in the hot molten liquor of previous process, run into vats, and recrystallized.—4th. The residuum in the dissolving tubs is smelted with lead and cupelled for its silver.

BY AMALGAMATION.

Patio Process.—1st. Patio signifies a yard. For amalgamating purposes, the floor of the yard is made level, paved with brick or granite blocks, surrounded by high walls and usually left open to the sky. On this floor, circular batches of silver ore, reduced to an impalpable paste by stamps and arastras, or other machinery, are spread to the depth of seven to twelve inches, and inclosed by low close curbs.—2d. Salt (chloride of sodium), varying in quantity according to its quality and the richness of the ore, is well mixed with the pulp by treading it with horses, mules or oxen, and turning it with shovels. The effect of the salt is to desulphurize the sulphides and produce chloride of silver. The batch is then left one entire day.—3d. Magistral, that is roasted and pulverized copper pyrites, varying in quantity with its quality, the richness of the ores and season, is well mixed with the pulp after it has been subjected to the treading and turning operation one hour. The ultimate effect of the magistral is to revive the silver by depriving it of chlorine.—4th. Quicksilver is added, usually in three charges to the mass, by being sprinkled in minute particles through cloth or other porous substance. After the addition of the first charge of quicksilver, the batch is thoroughly mixed, thrown into heaps of about one ton each, smoothed, and left at rest one whole day. The treading, turning and heaping operation is performed every other day, occupying five or six hours, and is found much more effective in a morning than an evening. The second charge of quicksilver is added and similarly treated, when it is ascertained by washing a small quantity of the mixture that the first has been well incorporated. After the second charge has performed its work, the third charge is added to take up any stray particles of silver, and to fit the amalgam better for separation.—5th. Lime is added to cool, and magistral to heat the mass, according as it may be too hot or too cold. Too much heat is indicated by the quicksilver becoming extremely divided and of a dark color, with occassional brown spots upon its surface; too little heat is indicated by the quicksilver retaining its natural color and fluidity. A proper degree of heat is indicated by the amalgam being of a grayish white color, and yielding readily to a slight pressure.—6th. The proportions to the ton of ore valued at fifty dollars are: salt, of good quality, 80 pounds; magistral (containing ten per cent. of the sulphate of copper), in summer, 20 pounds—in winter, 10 pounds; quicksilver (first charge) 14 pounds, (second charge) 5 pounds, (third charge) 7 pounds; lime, more or less (see Sec. 5), 15 pounds. An excess of magistral, quicksilver, or lime is injurious; an excess of salt causes a loss of quicksilver, but is not otherwise detrimental. The time employed in treating a batch of ore varies from twelve to sixty days. Light and good weather greatly facilitates operations.—7th. The separation is accomplished by washing the pulp or mixture with abundance of water in a large deep circular vessel, similar in principle to the common settler or separator, and causing the lighter portions of the mass to flow slowly off until the amalgam is gathered by itself, when it is taken to the refining works.

Freiberg Process.—1st. This process takes its name from Freiberg, a place in Germany where it was first practiced. The ores, if possible, are assorted so as to contain not less than twenty-five per cent. of sulphides. When they contain less, the sulphate of iron is added to make up the deficiency; when more, a sufficient quantity of the richest in sulphides is roasted without sea-salt to make good the ratio. The ores are crushed dry.—2d. Salt (chloride of sodium) and crushed ores are thoroughly mixed together, roasted in a reverberatory furnace, and reduced to an impalpable powder in a suitable mill. The heat and salt acting upon the sulphide of silver produce the chloride of silver.—3d. Wrought iron in small pieces with a pasty mixture of the chloridized ores and water, are put into barrels having horizontal axes, familiarly known as German barrels. These, each making twenty revolutions a minute, are run two hours. The effect of the iron is to revive the silver to the metallic state.—4th. Quicksilver is then poured into the barrels, after which they are run sixteen hours continuously, except the time taken to regulate the consistency of the pulp by the addition of ore or water. At the end of the time run, the casks are filled with water and revolved quite slowly for one or two hours, when the mass is discharged into large vats, and the amalgam separated by washing.—5th. The proportions to the ton of ore valued at seventy-five dollars per ton, are: salt, added before roasting, 200 pounds; wrought iron to the ton of roasted ore, 200 pounds; quicksilver to the ton of roasted ore, 1,000 pounds.

Ayer's Process.—The ore as it comes in blocks from the mines is heated in a furnace and thrown into brine or alkali water, by which operation it is broken into small fragments. It is then subjected to a further reduction and to the process of amalgamation.

Kent's Process.—"Kent's Patent Improved Freiberg Process, for treating gold and silver ores, consists in the preparation of crushed or pulverized ores, and tailings, by caking them with a solution of the chloride of sodium, or salt and water, introduced in any manner so as to admit of forming the crushed ore into cakes, lumps or bricks." Only three per cent. of salt is used. The bricks are calcined in a kiln, and it is estimated that three cords of wood and twelve hours' time will calcine 10,000 bricks, or 30 tons of ore. The amalgamation is effected in barrels—differing in some of their features from those of usual construction.

Veach Process.—The only essential difference between this and the Freiberg process consists in the employment of tubs instead of barrels, and the use of steam directly in the pulp. Vertical plates of iron or copper, for reviving the silver from the condition of a chloride, are fastened to the "muller arms," so as to revolve edgewise through the pulp or mass. The operations are greatly hastened by the application of steam, so that not more than five or six hours are required for the treatment of a charge of ore.

EXTRACTION OF SILVER FROM COARSE COPPER.

1st. Coarse copper containing silver, as it comes from the furnace, is raised first to a bright red heat, then pulverized and roasted with salt, by which operation the silver present is converted into the chloride of silver.—2d. The roasted mass is then treated in barrels with quicksilver, which decomposes the chloride and amalgamates with the revived silver in connection with a little copper and iron.—3d. The amalgam is retorted and cupelled.

EXTRACTION OF SILVER FROM COPPER MATTE.

1st. The copper matte, composed of the sulphides of silver, copper, iron and other base metals as it comes from the furnace, is pulverized and roasted with salt and lime. The former, by the action of its chlorine, converts the various metals into chlorides, which, with the exception of the chloride of silver, the latter decomposes.—2d. The roasted matte is then treated in barrels with quicksilver which decomposes the chloride of silver and amalgamates with the metal.—3d. The amalgam is then pressed, retorted and refined.

In Copper Kettles.—1st. This process, known as the "Hot Process," is chiefly employed in South America, and there only on the better class of ores—especially those rich in native silver, or in the chloride, iodide or bromide of silver. These being finely pulverized and concentrated by washing are introduced with considerable water into copper-bottomed kettles and boiled.—2d. Salt amounting by weight to ten or fifteen per cent. of that of the entire mass is dissolved in the boiling pulp while it is being stirred.—3d. Mercury of less weight than that of the silver present is added to the charge, and the stirring continued. More quicksilver is introduced whenever a test indicates the least dryness of amalgam.—4th. The amalgam and gangues are separated by washing; the former is "squeezed," retorted and refined, the latter often treated by the Patio process.

Pan Process.—1st. The ores of silver as they come from the mines are prepared for the pan usually by being crushed wet to a granular state by stamps and run into a series of large settling tanks. However, to crush *wet* and *very fine* at the same time is objectional, as much silver thereby is carried off in the water. The ores are commonly reduced so as to pass through No. 4 or No. 5 Screens. They are sometimes crushed dry, and sometimes roasted before going to the pan. Those especially containing much antimony should be roasted.— 2d. Charges of the reduced ores—raw for example—with sufficient water to form a soft, pasty mass, are put into the pans and ground to the slime condition. This ordinarily requires, in the Excelsior Grinder and Amalgamator about three hours. Water is added occasionally during the grinding process to preserve the proper consistence of the pulp, which, if too thick, causes a waste of power, and if too thin does not intimately mix with the quicksilver.—3d. Quicksilver is put into the machine and thoroughly incorporated with the mass; which condition is indicated by small globules of mercury appearing on the surface of the pulp while it is in motion, and by their slowly disappearing on its coming to rest. Millmen differ as to the proper time for its introduction; some maintaining that it should be simultaneous with the charging of the machine with ore, others that it should be later—that the grinding should first be accomplished, then the muller raised, the motion decreased, and the quicksilver mixed with the mass.—4th. Chem-

icals, differing in kind and proportions, to an indefinite
extent have been employed. As to their practical value,
it is sufficient to observe that, in treating the same
amount of ore, scarcely a pound is now used, while tons
formerly were. This observation applies to the treatment of raw ores containing the sulphide of silver. Prior
to drawing off the pulp, the pan may be nearly filled by
letting in water while the muller is running. This will
aid in depositing the amalgam and quicksilver, also in
drawing off the charge. The pan should be well washed
out before being recharged. When it is desired to
"clean up" the Excelsior Pan, the key in the muller-
screw must be removed, and the muller run up out of the
way of the operator. As the sides of the stationary
plates slant outward, it requires but very little labor to
wash down the material covering them to the circumference of the machine, whence it may readily be drawn
off.—5th. Heat nearly to the degree of boiling water is
universally approved of by the amalgamators of silver
in pans.—6th. The proportions usually observed in operating the "Excelsior Grinder and Amalgamator" with
muller four feet diameter, are: ore to the charge, 1,500
pounds; number charges worked in 24 hours, 7; revolutions of muller per minute, 60 to 65; quicksilver to
charge ore valued at fifty dollars a ton, 200 pounds; heat
(produced by steam), 200° Fahr.—7th. The charge is next
drawn off into the separator, where it is slowly and carefully washed in abundance of clean water. For the first
fifteen or twenty minutes after the separator has been

charged with the slime-pulp, but little more water than was drawn from the Grinder and Amalgamator or Pan should be employed in it, at the expiration of which time the machine may be nearly filled with water and kept running for a half hour or more; then the stoppers should be drawn from the side, one after the other, commencing at the uppermost, and the pulp run off slowly. In the meantime a clean stream of water should be let continuously into the machine. The Conoidal Separator being employed, the amalgam is deposited with the quieksilver in the bowl and spiral groove at the circumference. The most of the quicksilver is drawn off through an aperture in the bottom of the bowl. The amalgam is washed, squeezed and retorted.—Plumbiferous silver amalgam squeezed at a temperature of 144°–180° Fahr., the lead goes with the mercury; squeezed at a lower temperature, the lead remains in the bag. The pulp is run off from the separator into agitators and various contrivances employed to gather the stray particles of mercury, amalgam, and in some cases to concentrate the richer and heavier portions of the tailings.

RECIPES.—The following recipes taken from Guido Hüstel's "*Processes of Silver and Gold Extraction*," are a few of the many employed in Nevada, to the ton of ore:

a Chloride of copper................ 13 pounds
   Common salt..................... 60 "
b Chloride of iron................. 13 "

c Sulphate of iron.................... 1 pound

  Sulphate of copper................ 8 pounds

  Common salt...................... 80 "

d Sulphuric acid.................... 3 "

  Sulphate of copper................ 2 "

  Salt............................. 15 "

e Sulphuric acid.................... 2 "

  Alum............................ 2 "

  Sulphate of copper............... 1½ "

f Sulphate of copper .... 18 ounces

  Sulphate of iron................. 16 "

  Sal ammoniac.................... 8 "

  Common salt..................... 2 pounds

g Alum........................... 1½ "

  Sulphate of copper .... 1½ "

  Salt .... 40 "

h Muriatic acid................... 30 ounces

  Peroxide of manganese............ 8 "

  Blue vitriol..................... 10 "

  Green vitriol.................... 10 "

i Common salt.................... 15 pounds

  Nitric acid.................. 1 to 2 "

  Sulphate of iron............. 1 to 2 "

k Common salt .... 25 "

  Blue vitriol..... .... 2 "

  Catechu......................... 2 "

Note.—*a, b, c,* are calculated for ore containing from two hundred and fifty to five hundred ounces of silver in sulphurets. All chemicals, except salt, are used in solution. The salt is charged half an hour before the chemicals are put in.

SODIUM AMALGAM.

Sodium and mercury at a slightly elevated temperature combine energetically with each other, forming sodium amalgam, which has a strong affinity for gold and silver, and several other substances. The application of this principle of attraction to the practical amalgamation of gold and silver was first made by Prof. Henry Wertz of New York. Experience goes far to show that sodium amalgam, for the extraction of gold and silver from their gangues, in a variety of cases may by skillful hands be employed to advantage. A very free use, however, will often be found injurious on account of its affinity for many other substances than gold and silver. Its use is recommended in the treatment of gold-bearing quartz in batteries, pans, barrels or arastras, in extracting gold or silver from sweepings, and in amalgamating silver ores in which the silver has been reduced to the metallic state: also in cases of "flouring" and "sickening" of the quicksilver, and in "cleaning up," especially when a portion of the mercury and amalgam is in a finely divided state, or scum form. The ratio observed is: one part of sodium to two thousand or twenty-five hundred parts of quicksilver by weight.

7

PURIFICATION OF MERCURY.

Mercury, for the purposes of amalgamation, should be pure; most foreign substances, as lead, zinc, or bismuth, diminish its properties of combining with gold and silver. To free from these and other impurities:

1st. Distil the impure mercury. A retort for this process may readily be made of a common quicksilver flask and iron pipe of syphen form. The short leg of the pipe, a few inches long, is attached to the flask in the place of the removed stopper. The long leg, three or four feet in length, inclines downward from the bend. The retort should not be over two-thirds filled with mercury. The heat ought first to be applied to the short leg of the pipe and upper part of the retort, then to all parts of the flask alike. The long leg of the pipe must be kept cold. This may be effected by wrapping it with cloths and pouring on cold water. The discharge end may also be immersed in cold water, kept in the receiver. The heat should be uniform, and the distillation slow. The common covered retort is preferable to the one described.

2d. Heat, and frequently agitate the distilled mercury in thin sheets with one part nitric acid and two parts of pure water. The heat should be kept at 120° Fahr. for several hours. Repeat these operations until satisfactory results are obtained; then pour off the mercury for use.

3d. Digest the crystallized nitrate of mercury in nitric

acid; then dilute the solution, filter, precipitate the mercury by metallic copper, and add it to the mercury already obtained; or the nitrate of mercury may be converted to a liquid simply by heat, and the metal then precipitated by copper plate.

QUICKSILVERING OF COPPER-PLATE.

First, thoroughly cleanse the surface of the plate, and rub it over with quicksilver or with the nitrate of mercury. The surface is sometimes cleansed by simply scouring it with wood-ashes, brick-dust, or fine sand; and sometimes by washing it with dilute acid or strong alkali. When acid is employed, its corrosive properties should be neutralized before the application of the quicksilver. Nitrate of mercury, when crystallized, is readily converted to a liquid by heat, in which state it may be applied as a wash to the plate. Sodium amalgam may also be used to good effect in coating copper plate with quicksilver; in which case not as much care need be had in cleansing the surface of the plate as by the other methods, before its application.

CUPELLATION OF GOLD AND SILVER.

1st. The cupellation of gold and silver depends upon their permanence and the oxidability of lead with which in a state of fusion they readily enter into combination.

The mode of cupelling gold and silver in their several relations is the same.

2d. The alloy, consisting of lead and gold or silver, or lead, gold and silver, is melted in a circular reverberatory furnace, provided with openings through its sides for the admission of metal, heat, currents of air, and for the escape of vapors or litharge. The escape is opposite the blast opening. The roof, or top of the furnace, is of dome form, and movable. At each cupellation, the hearth, usually of concave form, is broken up and replaced by one made of clay, sand, and carbonate of lime.

3d. Blasts, or currents of air, are blown continually during the operation upon the surface of the fused alloy, promoting oxidation of the lead and causing the litharge or oxide of lead to pass out through the escape opening. The gate-way of this opening is kept level with the surface of the metal within. The metal thus separated from the lead remains on the hearth of the furnace, either as an alloy of gold and silver, or as one of them in nearly a pure state. It is deprived of what lead it may contain by the humid way of assay. If the metal obtained be an alloy of gold and silver, it is subjected to the process of inquartation, to which reference is here made.

REFINING OF GOLD AND SILVER.

ining of gold and silver is the process by which individually are brought to a state of purity.

It properly embraces Quartation, Granulation, Parting, and Reduction.

1st. *Quartation* is the alloying of one part of gold with three parts of silver, preparatory for humid operations.

2d. *Granulation* is effected by pouring the prepared melted alloy through a fine iron sieve into water, or directly upon a bundle of twigs immersed in water.

3d. *Parting* takes place by subjecting in a proper vessel, as glass, the granulated alloy to the action of boiling nitric acid, by which the silver with the base metals, if any are present, is dissolved out of the alloy, leaving the gold pure, and in the form of the original granules. In this operation 100 parts of silver require 149 parts of nitric acid of specific gravity 1,32 for oxidation and solution. Boiling concentrated sulphuric acid is sometimes employed instead of nitric acid.

4th. *Reduction.*—The granular gold is taken from the dissolving vessel, digested in boiling nitric acid, washed, dried, melted with a little nitre and cast into ingots. The silver of the nitric solution is precipitated by the introduction of copper plates. The precipitate is washed in water, pressed, melted with nitre, and borax, and cast into ingots. The silver in solution is sometimes converted into the chloride of silver by a solution of salt (chloride of sodium), and then reduced with carbonate of soda or other proper flux.

APPLIED MECHANICS.

THIN CYLINDERS.

To determine the thickness of thin hollow cylinders—the internal radius, pressure and tenacity of the material being given:

Rule.—Multiply the internal radius in inches by the fluid pressure in pounds per square inch, and divide the product by the tenacity per square inch of the material.

Ex.—The internal radius of a cylinder being thirty inches, the fluid pressure two hundred and fifty pounds to the square inch, and the tenacity of the material of the cylinder twelve thousand pounds per square inch, what is the thickness of the cylinder?

Cal. $250 \times 30 = 7500$; $7500 \div 12000 = \frac{5}{8}$ inch thick. Ans.

To determine the fluid pressure—the internal radius, thickness of cylinder, and tenacity of material being given:

Rule.—Divide the product of the thickness of the cylinder and tenacity of the material per square inch by the internal radius.

Ex.—The thickness of the cylinder being one-fourth of an inch, the tenacity eighteen thousand pounds, and the radius six inches, what fluid pressure will the cylinder stand per square inch?

Cal. $18000 \times \frac{1}{4} \div 6 = 750$ pounds.   Ans.

THICK HOLLOW CYLINDERS.

To determine the thickness of thick hollow cylinders —the internal radius, the fluid pressure, and the tenacity of the material of the cylinder being given:

Rule.—Subtract one (1) from the square root of the quotient of the sum and difference of the tenacity per square inch of the material of the cylinder and the fluid pressure per square inch, and multiply this difference by the internal radius.

Ex.—The internal radius of a thick hollow cylinder being nine inches, the tenacity of the material of the cylinder ten thousand pounds, and the fluid pressure eight thousand pounds per square inch; what is the requisite thickness of the cylinder?

Cal.—Sum of tenacity and fluid pressure, $10000 + 8000 = 18000$; difference of tenacity and fluid pressure, $10000 - 8000 = 2000$; quotient of sum and difference, $18000 \div 2000 = 9000$; square root of quotient, $\sqrt{9} = 3$; difference between root and one (1), $3 - 1 = 2$; product of radius and difference, $9 \times 2 = 18$ ins. Ans.

To determine the fluid pressure per square inch which a thick hollow cylinder will withstand—the internal and external radii and the tenacity of the material of the cylinder being given:

Rule.—Divide the difference of the squares of the radii, and multiply the quotient by the tenacity per square inch of the material of the cylinder.

Ex.—The internal and external radii of a thick hollow cylinder being respectively nine inches and twenty-seven inches, and the tenacity per square inch of the material of the cylinder ten thousand pounds, what fluid pressure per square inch will it withstand ?

Cal.—Square of external radius, $27 \times 27 = 729$; square of internal radius, $9 \times 9 = 81$; difference of squares, $729 - 81 = 648$; sum of squares, $129 + 81 = 810$; and $10000 \times 648 \div 810 = 8000$ pounds.   Ans.

RELATIVE GRINDING CAPACITY OF DIFFERENTLY FORMED GRINDING PLATES.

Rule 1.—To determine the relative grinding capacity of tractory-formed plates or mullers : Multiply the difference of the squares of the greater radius of the muller, and the radius of the opening in the same, by twice the greater radius.

Ex.—A tractory-muller having its greater radius one (1) and the radius of its opening one-third ($\frac{1}{3}$) what is its relative grinding capacity ?

Cal. $1 - \frac{1}{3} = \frac{8}{9}$; $\frac{8}{9} \times 2 = \frac{16}{9} = 1.778$.   Ans.

Rule 2.—To determine the relative grinding capacity of flat-bottomed or plane mullers : Multiply the difference of the squares of the greater radius of the muller, and the radius of the opening in the same, by the sum of the greater radius and the radius of the opening.

Ex.—A plane-muller having its greater radius one

and the radius of its opening one-third, what is its relative grinding capacity?

Cal. $1-\frac{1}{3}=\frac{2}{3}$; $1+\frac{1}{3}=\frac{4}{3}$; $\frac{2}{3}\times\frac{4}{3}=\frac{8}{9}=1.1852$. *Ans.*

Rule 3.—To determine the relative grinding capacity of conical plates or mullers—when the height of the cone is equal to one-half the greater radius: Find the grinding effect, as in Rule II. and multiply the result by 1.118. $1,1852\times1.118=1.3251$. *Ans.*

FRICTION.

1st. For a cylindrical journal, the leverage of the friction is simply the radius of the journal.—2d. For a flat pivot, the leverage is one-half the radius of the pivot.—3d. For a conical pivot the leverage is equal to the quotient arising by dividing one-half the radius of the base of the cone by the sine of the angle included between the axis and side of the cone.—4th. For Schide's "anti-friction" pivot the leverage is the entire radius of the pivot.—5th. For a collar, the leverage of friction is equal to one-half the sum of its external and internal radii.

WATER PIPES.

To determine the velocity of water per second, flowing through long pipes—the head or height of reservoir above the point of delivery, the length and diameter of the pipe being given:

Rule.—Multiply the product of the head and diameter

of the pipe in feet by twenty-three hundred: divide this product by the sum of once the length and fifty-two times the diameter of the pipe, and extract the square root of the quotient.

Ex.—The head is six hundred feet, the diameter of the pipe nine inches, the length of pipe six thousand feet— What is the velocity of the water per minute?

Cal.—Diameter of pipe=9 inches=.75 feet ; product of head, diameter, etc., $600 \times 2300 \times .75 = 1035000$; sum of length and product, $6000 + (52 \times .75) = 6039$; quotient, $1035000 \div 6039 = 171.386$; square root, $\sqrt{171.386} = 13.09$ feet velocity per second ; velocity per minute, $13.09 \times 60 = 785.4$ feet.   Ans.

To determine the head, the velocity, length and diameter of pipe being given—(by Tables V. and VI.) :

Rule.—Multiply the length of the pipe in feet by the constant (C) opposite the given velocity in Table, and divide the product by twelve hundred (1,200) times the sum of the diameter of the pipe in inches and the constant (C) opposite the given diameter of pipe in Table.

Ex.—It is required to determine what head of water is necessary to force water through 1,500 feet of six-inch pipe at a velocity of 180 feet a minute.

Cal.—By Table V., constant opposite velocity 180= 62.13 ; by Table VI., constant opposite diameter 6=.078 ; then $62.13 \times 1500 \div 1200 \ (6.078) = 12.77$ feet.   Ans.

To determine the velocity—the head, length and diameter of pipe being given—(by Tables V. and VI.) :

Rule.—Multiply twelve hundred (1,200) times the head

by the sum of the diameter of the pipe and its constant found in Table VI., and divide the product by the length of the pipe. The quotient will be the constant; opposite which find the velocity, in Table V.

Ex.—The head being 12.77 feet, the diameter 6 inches, and length of pipe 1,500 feet, what is the velocity?

Cal. 1200 × 12.77 × 6.078 ÷ 1500 = 62.13 constant, opposite constant 62.13 find velocity 180 feet. Ans.

The relative quantities of water, which will in the same time flow equal distances through straight, curved and right-angled pipes, are respectively 140, 100 and 90.

VELOCITY OF STREAMS.

In a stream the velocity is greatest at the surface and in the middle of the current. To find the velocity of a river or brook:

Rule.—Take the number of inches that a floating body passes over in one second in the middle of the current, and extract its square root; double this root, subtract it from the velocity at top and add one (1); the result will be the velocity of the stream at the bottom; and the mean velocity of the stream is equal the velocity at the surface, less the square root of the velocity at the surface increased by five-tenths of one (.5).

Ex.—If the velocity at the surface and in the middle of a stream be thirty-six inches a second, what is its velocity at the bottom, and what its mean velocity.

Cal. $\sqrt{36} \times 2 = 12$; then $36 + 1 - 12 = 25$ inches per second velocity at bottom. Ans. Then $36 + .5 - 6 = 30.5$ inches per second mean velocity. Ans.

WATER POWER.

The theoretical velocity with which a liquid issues from an orifice in the bottom or side of a vessel that is kept full, is equal to that which a heavy body would acquire by falling from the level of the surface to the level of the orifice; this is nearly eight times the square root of the head or distance fallen in feet. The practical velocity estimated for the entire opening is considerably less than the theoretical velocity, owing to oblique currents and to friction. The oblique currents produce a contraction in the vein or stream. The minimum transverse section of the contracted vein is the plane at which the velocity is nearly equal to the theoretical velocity. The quantity of water which will be discharged in a certain time depends upon the form of the opening as well as upon the head. Thus by means of a conical tube of the form of the contracted vein, the velocity at the opening or smaller end of the tube is nearly equal to the theoretical velocity. The actual velocity estimated for the entire opening when constructed as ordinarily and not large, is five and four-tenths the square root of the head in feet.

To determine the velocity with which water will flow

under different heads, and under different conditions, see Tables.

Ex. 1.—Let it be required to determine the velocity with which water will flow under a head of six inches through an aperture of ordinary construction.

Cal.—By Table IV. in the column called " Square root of head," and opposite the " Head six inches " we find, .70711. Multiply this number by 8, thus: .70711 × 8 = 5.66 feet per second; this is the theoretical velocity. Multiply the same number by 5.4; .70711 × 5.4 = 3.82 feet per second; this is the experimental velocity.

Ex. 2.—Let it be required to determine the number of pounds of water which will flow in one minute under a six-inch head through an aperture two inches by one hundred and fifty inches.

Cal.—The velocity per second, as found above is 3.82 feet. The velocity per minute then will be 3.82 × 60 = 229.10; the area of the opening is 150 × 2 = 300 square inches. This expressed in square feet, is 300 ÷ 144 = 2.08. Multiplying the length of the stream of water, or in other words the distance which the water flows in one minute, by the area of the opening, and we obtain the cubic feet of the quantity of water which will flow out in one minute, thus: 229.10 × 2.08 = 477.3 cubic feet. By Table XIII. we find that a cubic foot of water weighs 62.38 pounds; hence 477.3 × 62.38 = 29773.54 pounds.

Were it now required to find the power which this water would exert on a water-wheel, we should simply have to multiply the number of pounds of water thus found by the fall in feet; this would give us the theoretical power in the water. To determine the practical power, we should have to multiply the theoretical power by the efficiency of the water-wheel; and if we desired to express the power in units of horse-power, we should have to divide the pound power thus found by 33,000. As a practical illustration, take the common centre discharge-wheel, and let the fall be forty feet. The efficiency of this wheel under the most favorable circumstances does not exceed .60. Taking the highest efficiency, and we have (see Table I.) $29773.54 \times 40 \times .60 \div 33000 = 21.65$ horse-power.

To determine how much more water will flow under one head than under another :

Rule.—Divide the square root of the greater head in feet by the square root of the less.

Ex.—How much more water will flow under a three-foot head than under a six-inch head?

Cal.—By Table IV. we find the square root of the three-foot head to be 1.73205, and the square root of the six-inch head to be .70711. Then $1.73205 \div .70711 = 2.45$ times as much. Ans.

WATER-WHEELS.

Water, as a power or force, is exerted on water-wheels by its weight and by its impulse. Weight and impulse are combined on the overshot and breast-wheels. The theoretical work accomplished by weight is the product of its force and the vertical distance through which it is exerted. The theoretical work accomplished by impulse is the product of the force produced by the weight of the flow of water and the vertical height or head necessary to produce the velocity with which the weight moves.

The available work depends not only upon the magnitude of the force exerted, but upon the direction of that force in reference to the direction given to the resistance; also upon the form of the floats or buckets of the wheel, friction, losses by leakage, etc.

The velocity, per second, of the overshot wheel at its circumference should be about six feet; which is due a head of two feet three inches. The vertical distance from the centre of the opening in the gate to the surface of the water in the flume or reservoir is termed the "head," and the vertical distance from the centre of the opening in the gate to the lower edge of the wheel, the "fall."

To find the horse-power of various water-wheels:

Rule.—Multiply the product of the coefficient (see Table VII.) opposite the given head, the area of the opening in the gate in square inches, the entire head in feet (in case of the overshot or breast, the head by 40, and the fall by 78) by the efficiency of the class of wheel, pointing off six figures as decimals.

Ex. 1.—The dimensions of a stream are two inches by two hundred inches, the head two feet three inches, and the fall ten feet; what is its horse-power applied to a breast wheel?

Cal. 2 × 200 = 400 square inches gate opening; coefficient, by Table VII. opposite 2 feet 3 inches = 64; efficiency, by Table II. arising from impulse = 40; efficiency by Table II. arising from weight = 78; head 2 feet 3 inches = 2.25; product of efficiency and head, 2.25 × 40 = 90; product of efficiency and fall, 10 × 78 = 780; sum of products, 780 + 90 = 870; then 870 × 400 × 64 = 22.27 horsepower. Ans.

Ex. 2—The dimensions of the stream are ten inches square, the head twenty-five feet; what is its horsepower applied to a good turbine?

Cal.—Square inches in opening, 10 × 10 = 100; coefficient by Table VII. for 25 feet = 213; efficiency by Table II. of turbine, 68; then 100 × 213 × 68 × 25 = 36.21 horsepower. Ans.

STEAM POWER.

Steam, as a force, acts by elastic pressure. The law that, in compressing a perfect gas the volumes occupied by a given quantity are usually proportioned to the pressures, does not hold good in relation to saturated steam.

By Table VIII. it will be seen that the volume of steam under a pressure of thirty pounds to the square inch produced from a cubic inch of ice-cold water is eight hundred and thirty-eight cubic-inches, while under a pressure of ninety pounds to the square inch the volume is two hundred and ninety-eight cubic inches. Thus the ratio of the two pressures is as 30 to 90 or as 1 to 3, while the inverse ratio of the respective volumes of steam is as 1 to 2.81. The mechanical effect deduced from the above data is as follows : 838 × 30 ÷ 12=2095, and 298 × 90 ÷ 12=2235 ; 2235—2095=140, difference of mechanical effects, and 140 ÷ 2095=.0668 ; showing an advantage, all other things being equal, of nearly seven per cent. in favor of using steam at the higher pressure.

By Table IX. for estimating the mean pressure of steam for a given cut-off of stroke, the coefficient for one-fourth cut-off is 572 in the unjacketed cylinder, and 582 in the jacketed cylinder. The calculation on this data shows an advantage of one and three-fourths of one per cent. in favor of the jacketed cylinder for the given cut-off of stroke.

The back pressure of steam in the cylinder of an engine of ordinary structure is found by experience to be about four pounds to the square inch above the atmospheric pressure ; the velocity of piston being three hundred feet per minute. But in the best engines of modern construction, the back pressure is frequently below one pound, with even higher speeds of piston. It is also found that the excess of the back pressure above the atmospheric pressure varies nearly as the velocity of the piston.

To find the mean pressure of steam for a given cut-off of stroke :

Rule.—Multiply the excess of the pressure of steam above the atmospheric pressure, per square inch, as it enters the cylinder, by the tabular coefficient opposite the given cut-off, pointing off three figures as decimals, and deduct from the product the tabular correction for the same cut-off.

Ex.—With steam entering the cylinder at a pressure of ninety pounds to the square inch, and cut off at three-tenths ($\frac{3}{10}$) stroke, what is the mean pressure?

Cal.—For unjacketed cylinder. By Table IX. the co-efficient for $\frac{3}{10}$ stroke=.639. Correction for same=5.307. Then .639 × 90=57.510; 57.510—5.307=52.203 pounds. Ans.

Cal.—For jacketed cylinder. By Table IX. the coefficient for $\frac{3}{10}$ stroke is=.648. Correction for same=5.174. Then, 648 × 90=58.320; 58.320—5.174=53.146 pounds. Ans.

The "indicated" power of a steam-engine is equal to the product of the area of the piston in inches, the pounds pressure of the steam per square inch, and the distance travelled by the piston in one minute. The effective power in excess of back pressure, loss and condensation in well-constructed engines is usually about eighty-four (.84) one hundredth of the indicated or theoretical power. The nominal power bears no fixed relation to the indicated, nor to the effective power. In calculating the nominal power, the mean effective pressure in low pressure engines is assumed to be seven (7) pounds, and in high pressure engines twenty-one (21) pounds. The nominal velocity of the piston, whether in high or low pressure engines, is 128 times the cube root of the stroke in feet. The range is from 120 to 300 feet per minute. In practice the piston not unfrequently is made to travel

900 feet and upwards a minute. Experience has determined no limit beyond which it cannot be run with advantage.

The total heat of combustion estimated in pounds of water raised 1° Fahr. is 14,500 times the weight of fuel —pure carbon. This is equivalent to 11,194,000 foot pounds of work per pound of fuel. It is also equivalent to 966 times the weight of water which the same fuel will evaporate from 212° Fahr. In practice, not usually over six per cent. of this immense expenditure of heat is economized. From five to seven pounds of coal, or from 20 to 30 pounds of pine wood, ordinarily are allowed to the horse-power per hour. But in the best examples of modern practice a horse-power per hour, with an expenditure of from three to four pounds of good coal, is not unfrequent. The Babcock & Wilcox engine in the Druid Mill, near Baltimore, develops power for 2.24 pounds of chestnut coal per horse-power per hour, which is probably the best result yet attained in stationary engines. It may further be remarked that in practice, "for each nominal horse-power, a boiler requires one cubic foot of water per hour, one square foot of fire-grate surface, one square yard of heating surface, one cube yard capacity, twenty-eight inches flue area." Or for each nominal horse-power a tubular boiler, working to about double its nominal horse-power, requires one cubic foot of water per hour, ten square feet of heating surface (the whole tube surface being taken as effective), one-half square foot of fire-grate, ten square inches sectional area of tubes, thirteen square inches flue area, six square inches chimney area, eight cube feet total boiler capacity and two cube feet of steam room.

The diameter of two cylinders being the same, for instance, six inches each, and the length of stroke of one ten inches, and of the other fourteen inches, the power of each, theoretically, will be the same, the velocity of piston being the same. In practice, there is a slight, and but a slight, percentage in favor of the longer cylinder. In two cylinders, one being ten inches and the other twelve inches in diameter, and each twenty-four inches long, and their pistons run under the same pressures per square inch, and at the same velocities, their inside relative friction would be as 25 to 30, and aside from this small difference, their relative powers would be as 25 to 36. Again, in two cylinders, one sixteen inches and the other twenty inches diameter, each forty-eight inches long, run under similar circumstances, the inside relative friction will be as 100 to 125, while their respective powers will be as 100 to $156\frac{1}{4}$. This small difference in friction, as compared to the whole, is not generally taken into account.

To find the effective horse-power of a non-condensing steam-engine:

Rule.—Multiply four times the square of the diameter of the piston in inches by the product of the number of revolutions, length of stroke in feet, and the average forward pressure of steam in pounds per square inch above atmosphere, pointing off five figures as decimals.

Ex.—What is the effective horse-power of an engine, the diameter of the piston being sixteen inches, the length of stroke three feet, the number of revolutions fifty per minute, and the average forward pressure, above the atmospheric pressure, seventy-one pounds per square inch.

Cal. $16 \times 16 \times 4 \times 50 \times 3 \times 71 = 109.05.$

Ex. 2.—What is the effective horse-power of an engine, the diameter of the piston being twelve inches, the length of stroke two feet, the pressure of steam as it enters the cylinder sixty pounds in excess of atmospheric pressure, cut-off at one-half stroke, and making seventy-five revolutions per minute?

By Table IX. coefficient for ½ stroke = .833; correction for same = 2.455; .833 × 60 = 49.980, 49.980 = 2.455 = 47.525 average pressure. Then $12 \times 12 \times 4 \times 47.525 \times 2 \times 75 = 38.59$ horse-power. Ans.

NITRO-GLYCERINE.

Nitro-Glycerine has for its formula $C_4 N_3 H_5 O_6$, which by composition is carbon 17.06, nitrogen 19.91, hydrogen 2.36, oxygen 60.66.

Preparation.—Syrup of Glycerine is slowly dropped into a mixture composed of equal volumes of nitric and sulphuric acids contained in a strong vessel surrounded by ice. The mixture is thoroughly agitated after each addition of the syrup. The nitro-glycerine of an oily form is removed from the surface and repeatedly washed with clean water. It is sometimes dissolved in alcohol or ether, and precipitated with water. The only precaution necessary in its preparation is to keep it cold.

Properties.—It is a pale yellow liquid, heavier than water, inodorous when pure; has a sweet, pungent and aromatic taste. It is poisonous, producing headache when placed even in small quantities upon the tongue, or when its vapors are inhaled. Its explosive power is about ten times that of ordinary blasting powder. It explodes when pure and at a moderate temperature by concussion alone. It is soluble in pyroligneous acid, and when in solution is not explosive. Water precipitates it from its solution.

Advantages.—1st. It requires a smaller hole than gun-powder.—2d. It obviates tamping, as water is employed to fill the holes.—3d. If used extensively it would be cheaper than gunpowder.—4th. It may readily be exploded in water.—5th. It can easily be manufactured at the mine or place where it is to be used.

Disadvantages.—1st. It (when pure) explodes by concussion alone.—2d. It (when impure) is liable to explode spontaneously.—3d. It explodes at a temperature of 300° Fahr.—4th. It is poisonous, as also are its vapors, thus rendering its use impracticable in narrow subterranean workings, especially if they are not well ventilated. However, when it is better understood these disadvantages may be removed.

TABLES.

TABLE I.

Showing the best velocities of various water-wheels as compared with the supply velocity :

Undershot and low breast, at circumference .50

Turbines, at the middle of ring of buckets.. .65

Reaction, at circumference................ .97

Central discharge (common).............. .60

Overshot, at circumference................ .50

TABLE II.

Showing the average efficiency of various water-wheels running under favorable circumstances as found by experience:

Undershot, having flat radial floats........ .35

Poncelet, improved undershot............. .60

Turbine (for example Jonval)............. .68

Reaction (for example the Scotch Turbine). .66

Central discharge (common).............. .55

Overshot and Breast (that part of fall acting by weight)...................... .78

Overshot and Breast (that part of fall acting by impulse)..................... .40

TABLE III.

Showing the relative velocities with which water, under a constant head, will flow through differently formed apertures:

Theoretical velocity....................=8.

Velocity through a thin plate..........=5.

Velocity through a tube two or three diameters in length projected outward=6.5

Velocity through a tube of the same length projected inward............=5.45

Velocity through a conical tube of the form of the contracted vein........=7.9

TABLE IV.

For estimating the velocity with which water will flow under a given head, or a body fall from a given height:

Head.	Sq. Root of Head.	Head.	Sq. Root of Head.	Head.	Sq. Root of Head.	Head.	Sq. Root of Head.
ft. in.		ft. in.		ft.		ft.	
1	.28867	2 4	1.52752	5	2.23607	32	5.65685
2	.40825	2 5	1.53456	6	2.44949	33	5.74456
3	.50000	2 6	1.58114	7	2.64575	34	5.83095
	.57735	2 7	1.60727	8	2.82843	35	5.91608
	.64549	2 8	1.63299	9	3.00000	36	6.00000
	.70711	2 9	1.65831	10	3.16228	37	6.08276
	.76376	2 10	1.68325	11	3.31662	38	6.16441
	.81744	2 11	1.70782	12	3.46410	39	6.24500
	.6602	3 0	1.73205	13	3.60556	40	6.32456
	.187	3 1	1.75589	14	3.74166	45	6.70820
	3	3 2	1.77952	15	3.87298	50	7.07107

Table IV. continued.

Head. ft. in.	Sq. Root of Head.	Head. ft. in.	Sq. Root of Head.	Head. ft.	Sq. Root of Head.	Head. ft.	Sq. Root of Head.
I 0	1.00000	3 3	1.80278	16	4.00000	55	7.41620
I I	1.04084	3 4	1.82574	17	4.12311	60	7.74597
I 2	1.08012	3 5	1.84842	18	4.24264	70	8.36660
I 3	1.11803	3 6	1.87082	19	4.35890	80	8.94427
I 4	1.15469	3 7	1.89297	20	4.47214	90	9.48383
I 5	1.19024	3 8	1.91485	21	4.58258	100	10.00000
I 6	1.22474	3 9	1.93649	22	4.69042	125	11.18034
I 7	1.25830	3 10	1.95789	23	4.79583	150	12.24745
I 8	1.29100	3 11	1.97909	24	4.89898	175	13.22876
I 9	1.32288	4 0	2.00000	25	5.00000	200	14.14214
I 10	1.35400	4 1	2.02072	26	5.09902	225	15.00000
I 11	1.38444	4 2	2.04124	27	5.19615	250	15.81139
2 0	1.41421	4 3	2.06156	28	5.29150	300	17.32051
2 I	1.44339	4 4	2.08167	29	5.38376	400	20.00000
2 2	1.47196	4 5	2.10159	30	5.47723	500	22.36068
2 3	1.50000	4 6	2.12132	31	5.56776	600	24.49490

TABLE V.

Showing the heads of water necessary to maintain different velocities of water in one hundred feet of pipe:

V.	C.	V.	C.	V.	C.
60	8.62	90	17.95	140	38.90
70	11.40	100	21.56	150	44.00
80	14.58	120	29.70	180	62.13

V. represents the velocities in feet per minute. C represents the constant numbers for these velocities.

8

TABLE VI.

Showing the constant numbers for different diameters:

D.	C.	D.	C.	D.	C.
4	.028	6	.078	8	.134
5	.053	7	.104		

D. represents the diameter of the pipe in inches. C. represents the constant numbers for the diameters.

TABLE VII.

Coefficients for estimating the horse-power of water-wheels:

Head ft.	Head in.	Coefficient	Head ft.	Head in.	Coefficient	Head ft.	Head in.	Coefficient	Head ft.	Head in.	Coefficient
	1	12	1	7	54	3	2	76	9	0	128
	2	17	1	8	55	3	4	78	10	0	135
	3	21	1	9	56	3	6	80	12	0	148
		25	1	10	58	3	8	82	14	0	160
			1	11	59	3	10	84	16	0	171
			2	0	60	4	0	85	20	0	191
		33	2	1	62	4	3	88	25	0	213
		35	2	2	63	4	6	90	30	0	233
	9	37	2	3	64	4	9	93	36	0	256
	10	39	2	4	65	5	0	95	49	0	298
	11	41	2	5	66	5	4	98	64	0	341
	0	43	2	6	67	5	8	101	81	0	384
	1	44	2	7	69	6	0	104	100	0	426
	2	46	2	8	70	6	6	109	121	0	469
	3	48	2	9	71	7	0	113	144	0	511
	4	49	2	10	72	7	6	117	169	0	554
		51	2	11	73	8	0	121	196	0	597
		52	3	0	74	8	6	124	225	0	639

TABLE VIII.

Showing the pressures, temperatures and volumes of steam.

P.	T.	V.	P.	T.	V.	P.	T.	V.
1.	102.1°	20582	60	292.7°	437	180	372.9°	155
5.	162.3	3813	65	298.0	405	190	377.5	148
10.	193.3	2358	70	302.9	378	200	381.7	141
14.7	212.0	1642	75	307.5	353	210	386.0	135
15.	213.1	1610	80	312.0	333	220	389.9	129
20.	228.0	1229	90	320.2	298	230	393.8	123
25.	240.1	996	100	327.9	270	240	397.5	119
30.	250.4	838	110	334.6	247	250	401.1	114
35.	259.3	726	120	341.1	227	260	404.5	110
40.	267.3	640	135	350.1	203	270	407.9	106
45.	274.4	572	150	358.3	184	280	411.2	102
50.	281.0	518	165	366.0	169	300	417.5	96

REMARK.—In the above table, P. denotes the total pressure in pounds per square inch; T. the corresponding temperature, Fahr.; and V. the volume of the steam compared to the volume of the water that has produced it.

TABLE IX.

For estimating the mean pressure of steam for a given cut-off of stroke:

	UNJACKETED CYLINDER.			JACKETED CYLINDER.	
Cut-off.	Coefficient.	Correction.	Cut-off.	Coefficient.	Correction.
$\frac{1}{20}$	.177	12.098	$\frac{1}{20}$	.186	11.966
$\frac{3}{40}$	.244	11.113	$\frac{3}{20}$	.254	10.966
$\frac{1}{10}$	.303	10.246	$\frac{1}{10}$	.314	10.084
$\frac{1}{8}$	.356	9.467	$\frac{1}{8}$	.370	9.261
$\frac{3}{20}$	.407	8.717	$\frac{3}{20}$	.417	8.570
$\frac{1}{5}$	.496	7.409	$\frac{1}{5}$	.505	7.297
$\frac{1}{4}$	.572	6.290	$\frac{1}{4}$	.582	6.145
$\frac{3}{10}$	.639	5.307	$\frac{3}{10}$	.648	5.174
$\frac{7}{20}$	.697	4.454	$\frac{7}{20}$	.707	4.307
$\frac{2}{5}$	.748	3.704	$\frac{2}{5}$	.756	3.587
$\frac{9}{20}$	.797	2.984	$\frac{9}{20}$	.800	2.940
$\frac{1}{2}$	.833	2.455	$\frac{1}{2}$	.840	2.352
$\frac{11}{20}$	.869	1.926	$\frac{11}{20}$	.874	1.852
$\frac{3}{5}$	.894	1.558	$\frac{3}{5}$	.900	1.470
$\frac{13}{20}$	.923	1.132	$\frac{13}{20}$	.929	1.044
$\frac{7}{10}$	.945	0.808	$\frac{7}{10}$	.945	0.808
$\frac{3}{4}$	.960	0.588	$\frac{3}{4}$	.960	0.588
$\frac{4}{5}$	.976	0.353	$\frac{4}{5}$	.976	0.353
$\frac{17}{20}$	.986	0.206	$\frac{17}{20}$	.986	0.206
$\frac{9}{10}$	.997	0.044	$\frac{9}{10}$	.997	0.044

TABLE X.

SCOUR OF WATER-COURSE BEDS.

$\frac{1}{4}$ foot per second will scour fine clay.

$\frac{1}{2}$ " " " " " sand.

$\frac{3}{4}$ " " " " coarse sand.

1 " " " " fine gravel.

2 feet " " " round shingle one inch diam.

3 " " " " augular stones, size of an egg.

5 " " " " conglomerate.

TABLE XI.

MEASURES OF ROCK, EARTH, ETC.

25 cubic feet of sand $=1$ ton.

18 cubic feet of earth $=1$ ton.

17 cubic feet of clay $=1$ ton.

13 cubic feet of quartz, unbroken in lode $=1$ ton.

18 cubic feet of gravel or earth, before digging $=27$ cubic feet when dug.

20 cubic feet of quartz, broken (of ordinary fineness coming from the lode) $=1$ ton contract measurement.

TABLE XII.

HARDNESS OF MINERALS—MAP-SCALE.

1. Talc; common laminated, light-green variety.
2. Gypsum; a crystallized variety.
3. Calcareous spar; transparent variety.

4. Fluor-spar; crystalline variety,

5. Apatite; transparent variety.

5, 5 Scapolite; crystalline variety.

6. Feldspar; white cleavable variety.

7. Quartz; transparent.

8. Topaz; transparent.

9. Sapphire; cleavable varieties.

10. Diamond.

Note.—Care must be taken to apply the file to edges of equal obtuseness.

MISCELLANEOUS XIII.

5760 grains=1 pound troy=1 pound apothecary

480 grains=1 ounce troy=1 ounce apothecary.

12 ounces=1 pound troy=1 pound apothecary.

7000 grains=1 pound avoirdupois.

437.5 grains=1 ounce avoirdupois.

16 ounces avoirdupois=1 pound avoirdupois.

252.458 grains=1 cubic inch distilled water, English standard 62° Fahr., barometer at 30 inches.

252.693 grains=1 cubic inch distilled water, U. S. standard 39.83° Fahr., barometer at 30 inches.

27.7015 cubic inches distilled water=1 pound avoirdupois.

1 cubic foot distilled water=62.37929 pounds avoirdupois.

REMARK.—In ordinary calculations a cubic foot of fresh water is taken at 62.5 pounds avoirdupois.

1 cubic foot of salt or sea water=64.3 pounds avoirdupois.

231 cubic inches=8.388822 pounds avoirdupois=1 gallon U. S. standard.

277.274 cubic inches=10 pounds avoirdupois=1 gallon imperial.

221.184 cubic inches=8 pounds of pure water at its maximum density=1 gallon N. Y. statute measure.

2150.42 cubic inches=77.627413 pounds avoirdupois= 1 bushel U. S. standard.

REMARK.—The dimensions of the bushel measure are 18.5 inches diameter inside, 19.5 inches outside, and 8 inches deep; and when heaped the cone must not be less than 6 inches high, equal to 2747.7 cubic inches for a true cone.

2211.84 cubic inches=80 pounds of pure water at its maximum density=1 bushel State of New York standard.

1 grain gold 1,000 fine	=	$ .0430663 mint value.
1 grain silver "	=	.0026936 "
1 grain copper "	=	.0000595 "
1 ounce gold "	=	20.671791 "
1 ounce silver "	=	1.292929 "
1 ounce copper "	=	.028571 "

REMARK.—Gold and silver, pure, are said to be 1,000 fine, or 24 carats fine.

The standard fineness of the United States coin is 900, or $24 \times .900 = 21.6$ carats fine.

www.ingramcontent.com/pod-product-compliance
Lightning Source LLC
Chambersburg PA
CBHW080656190526

45169CB00006B/2137